Intermediate 1 | Units 1, 2 & Applications

Mathematics

2001 Exam
Paper 1 (Non-calculator)
Paper 2

2002 Exam
Paper 1 (Non-calculator)
Paper 2

2003 Exam
Paper 1 (Non-calculator)
Paper 2

2004 Exam
Paper 1 (Non-calculator)
Paper 2

2005 Exam
Paper 1 (Non-calculator)
Paper 2

First exam published in 2001.
Published by Leckie & Leckie, 8 Whitehill Terrace, St. Andrews, Scotland KY16 8RN tel: 01334 475656 fax: 01334 477392
enquiries@leckieandleckie.co.uk www.leckieandleckie.co.uk

ISBN 1-84372-320-4

A CIP Catalogue record for this book is available from the British Library.

Printed in Scotland by Scotprint.

Leckie & Leckie is a division of Granada Learning Limited, part of ITV plc.

Acknowledgements

Leckie & Leckie is grateful to the copyright holders, as credited at the back of the book, for permission to use their material.
Every effort has been made to trace the copyright holders and to obtain their permission for the use of copyright material.
Leckie & Leckie will gladly receive information enabling them to rectify any error or omission in subsequent editions.

[BLANK PAGE]

FOR OFFICIAL USE

Total mark

X056/102

NATIONAL QUALIFICATIONS 2001

THURSDAY, 17 MAY
9.00 AM – 9.35 AM

MATHEMATICS
INTERMEDIATE 1
Units 1, 2 and Applications of Mathematics
Paper 1
(Non-calculator)

Fill in these boxes and read what is printed below.

Full name of centre

Town

Forename(s)

Surname

Date of birth
Day Month Year

Scottish candidate number

Number of seat

1 **You may NOT use a calculator.**

2 Write your working and answers in the spaces provided. Additional space is provided at the end of this question-answer book for use if required. If you use this space, write clearly the number of the question involved.

3 Full credit will be given only where the solution contains appropriate working.

4 Before leaving the examination room you must give this book to the invigilator. If you do not you may lose all the marks for this paper.

SCOTTISH QUALIFICATIONS AUTHORITY

FORMULAE LIST

Circumference of a circle:	$C = \pi d$
Area of a circle:	$A = \pi r^2$
Curved surface area of a cylinder:	$A = 2\pi rh$

Theorem of Pythagoras:

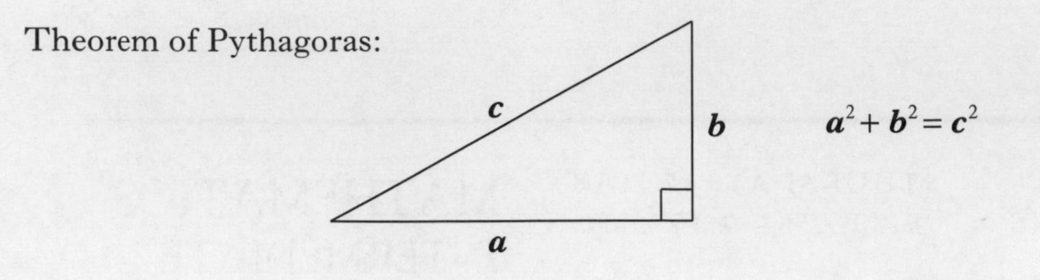

$$a^2 + b^2 = c^2$$

DO NOT WRITE IN THIS MARGIN

Marks

ALL questions should be attempted.

1. (*a*) Find $7{\cdot}35 \times 8$. **1**

(*b*) Find $\frac{3}{4}$ of £82. **1**

2. Part of the timetable of the overnight bus from Stirling to London is shown opposite.

Stirling (depart)	2140
London (arrive)	0615

How long does the journey from Stirling to London take? **1**

[Turn over

DO NOT WRITE IN THIS MARGIN

Marks

3. Eight jars of jam can be made from 2 kilograms of raspberries.
How many jars of jam can be made from 5 kilograms of raspberries?

2

4. Jenna is buying a car. The cash price is £11 500. It can be bought on hire purchase by paying a deposit of 20% of the cash price and 36 instalments of £300.

Find the total hire purchase price of the car.

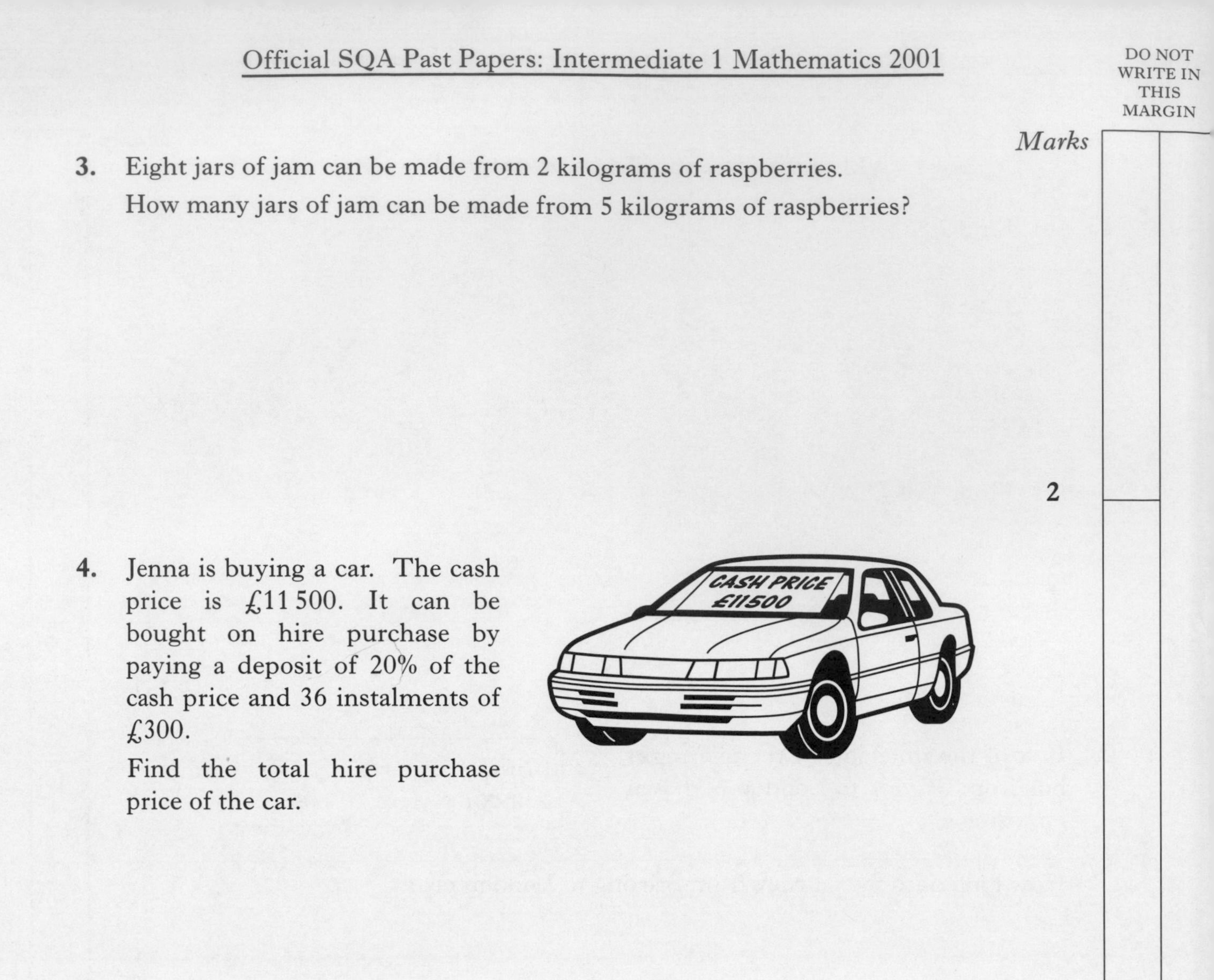

3

DO NOT WRITE IN THIS MARGIN

Marks

5. This network diagram shows the distances (in miles) between five towns.

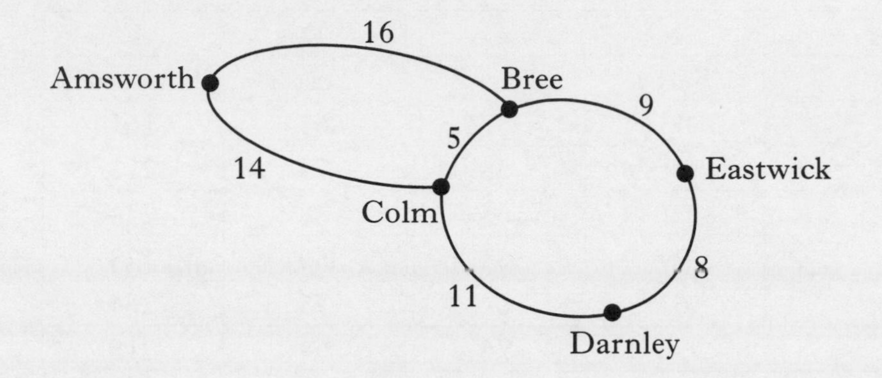

One route which starts at Amsworth and passes through the other four towns **once** is

Amsworth → Bree → Eastwick → Darnley → Colm

Distance = 16 + 9 + 8 + 11 = 44 miles

Write down all other possible routes which start at Amsworth and pass through the other four towns once.

Find the shortest route.

3

[Turn over

DO NOT WRITE IN THIS MARGIN

Marks

6. (*a*) This spreadsheet is used to record students' marks.

	A	B	C	D	E
1		Test 1	Test 2	Test 3	Total mark
2	James	43	57	61	161
3	Ann	81	56	74	211
4	Sarah	71	65	88	224
5	William	29	35	59	123
6	Ian	63	66	75	204
7	Mhairi	75	76	64	215
8					
9					

The formula =SUM(B4..D4) was used in this spreadsheet.

In which cell was this formula used?

1

(*b*) Another spreadsheet looks like this.

	A	B	C	D	E	F
1		Test 1	Test 2	Test 3	Test 4	Average mark
2	Sadik	56	63	67	57	
3	Ronald	64	68	69	59	65.0
4	Gail	71	59	72	64	66.5
5	Norma	69	71	58	66	66.0
6	Peter	74	62	69	71	69.0
7						
8						
9						

What formula is used in cell F2 to work out Sadik's average mark?

1

DO NOT WRITE IN THIS MARGIN

Marks

7. During a period of 30 days the temperature at a weather station is recorded each day.

The frequency table below shows these temperatures.

Temperature (°C)	*Frequency*	*Temperature* × *Frequency*
–3	1	
–2	2	
–1	4	
0	2	
+1	6	
+2	8	
+3	3	
+4	4	

(*a*) Write down the modal temperature. **1**

(*b*) Complete the table above and find the mean temperature.
Give your answer as a decimal. **3**

[Turn over

DO NOT WRITE IN THIS MARGIN

Marks

8. The diagram below shows the net of a triangular prism.

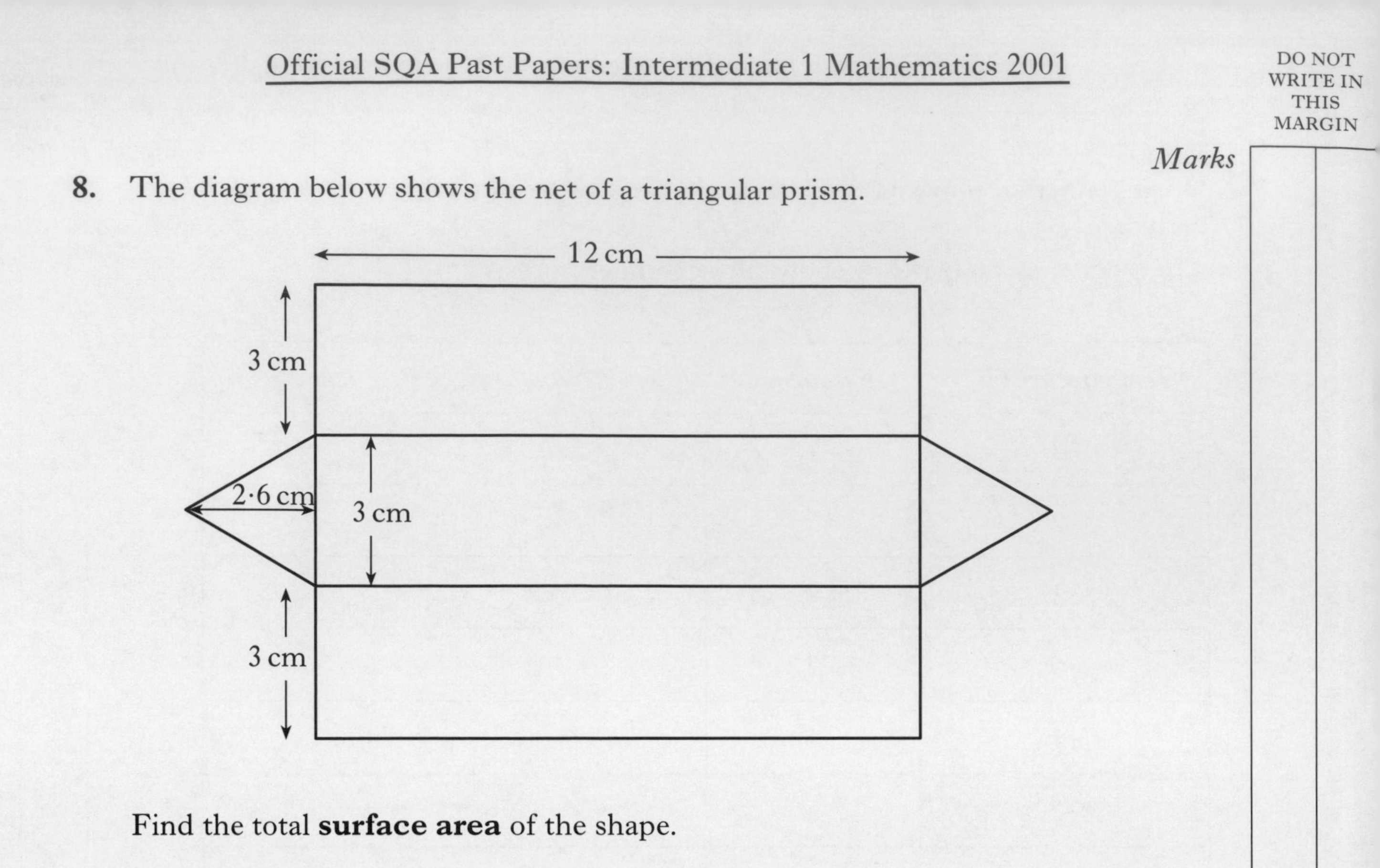

Find the total **surface area** of the shape.

4

DO NOT WRITE IN THIS MARGIN

Marks

9. The manager of the Central Hotel is buying new televisions for each of the hotel's 50 bedrooms. Two suppliers offer him the following deals.

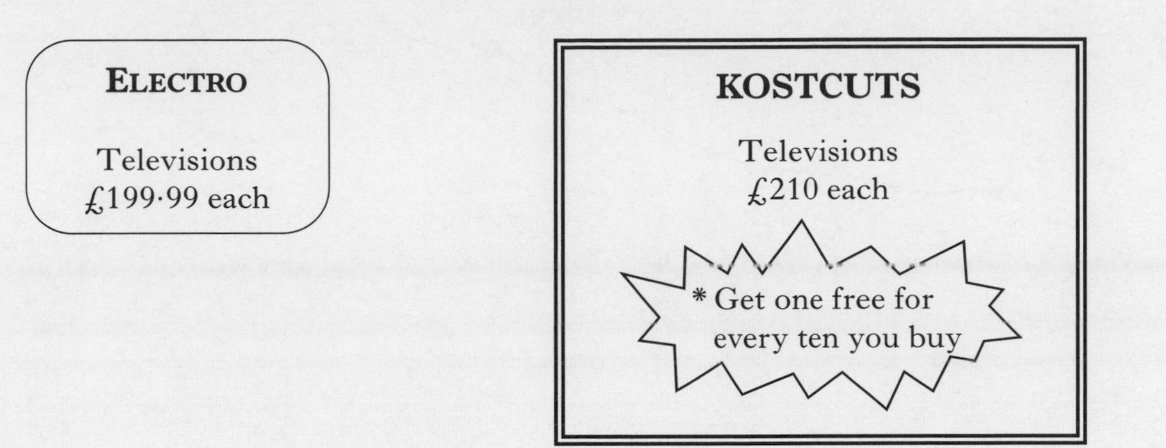

Which supplier offers the lower price for 50 televisions?

You must show your working.

4

[Turn over

DO NOT WRITE IN THIS MARGIN

Marks

10. The scale drawing shows the position of two ports, Lubness and Moy.

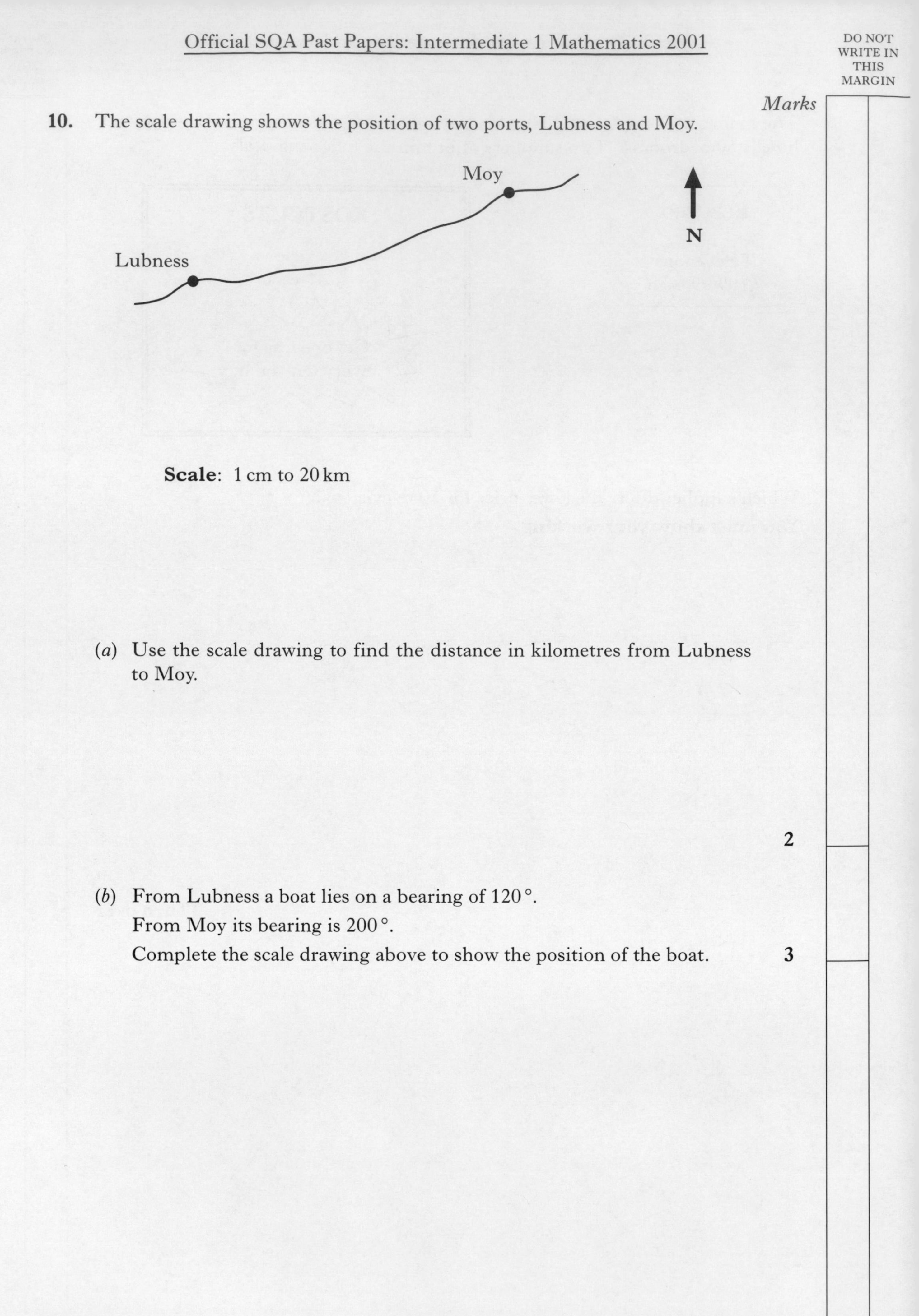

Scale: 1 cm to 20 km

(*a*) Use the scale drawing to find the distance in kilometres from Lubness to Moy. **2**

(*b*) From Lubness a boat lies on a bearing of 120 °.
From Moy its bearing is 200 °.
Complete the scale drawing above to show the position of the boat. **3**

DO NOT WRITE IN THIS MARGIN

Marks

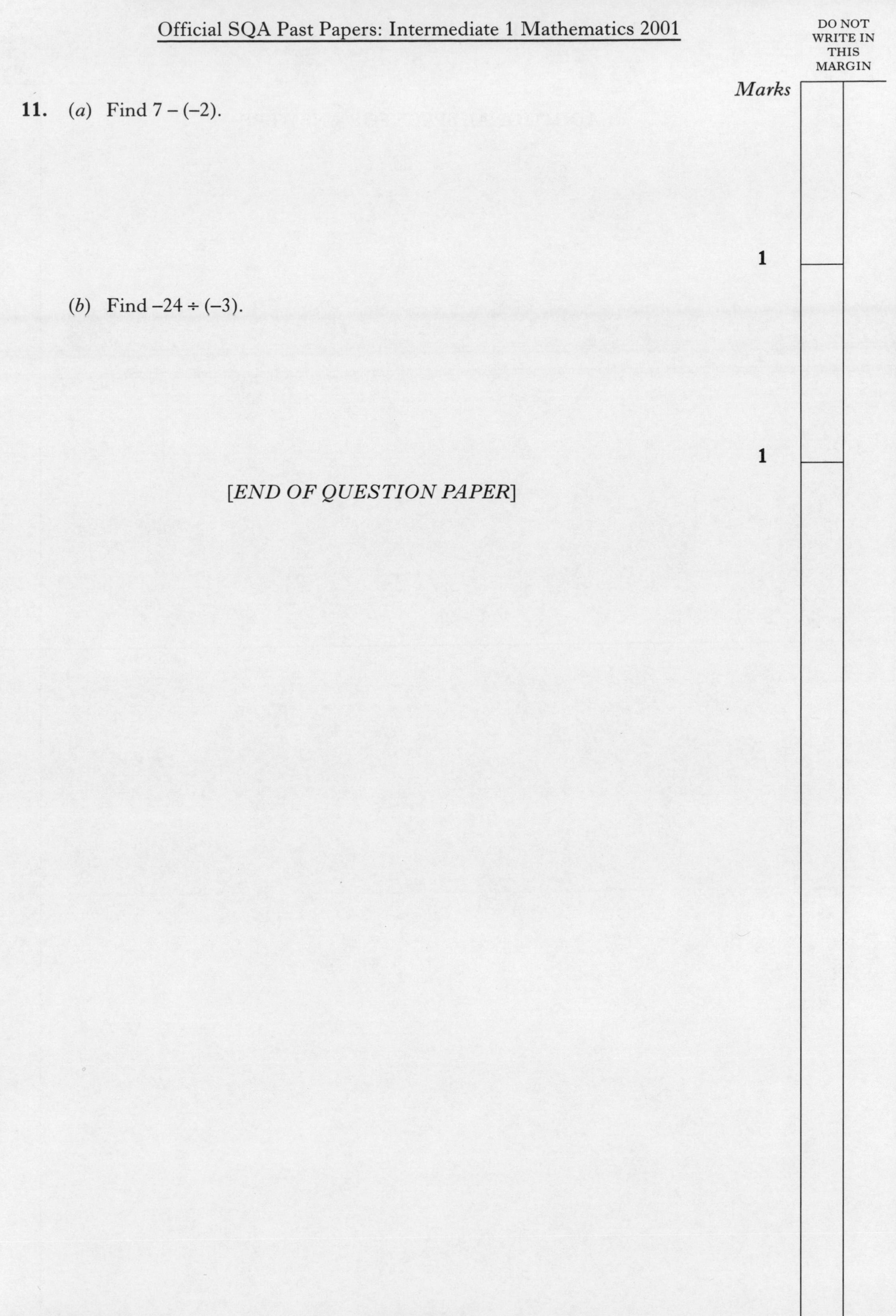

11. (*a*) Find $7 - (-2)$. **1**

(*b*) Find $-24 \div (-3)$. **1**

[*END OF QUESTION PAPER*]

DO NOT WRITE IN THIS MARGIN

ADDITIONAL SPACE FOR ANSWERS

FOR OFFICIAL USE

Total marks

X056/104

NATIONAL QUALIFICATIONS 2001

THURSDAY, 17 MAY
9.55 AM – 10.50 AM

MATHEMATICS
INTERMEDIATE 1
Units 1, 2 and
Applications of Mathematics
Paper 2

Fill in these boxes and read what is printed below.

Full name of centre

Town

Forename(s)

Surname

Date of birth
Day Month Year

Scottish candidate number

Number of seat

1 **You may use a calculator.**

2 Write your working and answers in the spaces provided. Additional space is provided at the end of this question-answer book for use if required. If you use this space, write clearly the number of the question involved.

3 Full credit will be given only where the solution contains appropriate working.

4 Before leaving the examination room you must give this book to the invigilator. If you do not you may lose all the marks for this paper.

LIB X056/104 6/12270

FORMULAE LIST

Circumference of a circle:	$C = \pi d$
Area of a circle:	$A = \pi r^2$
Curved surface area of a cylinder:	$A = 2\pi rh$

Theorem of Pythagoras:

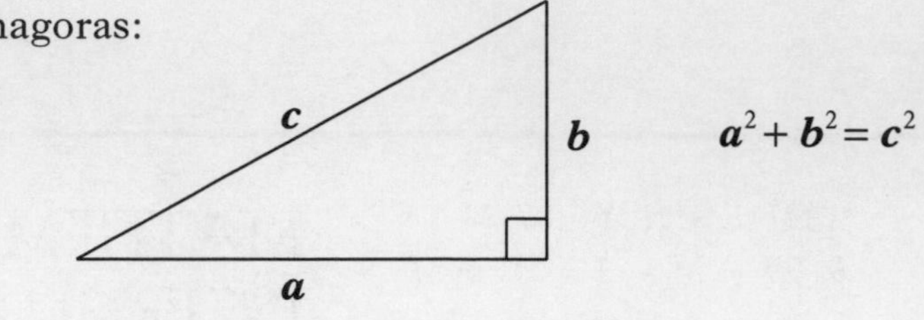

$a^2 + b^2 = c^2$

DO NOT WRITE IN THIS MARGIN

Marks

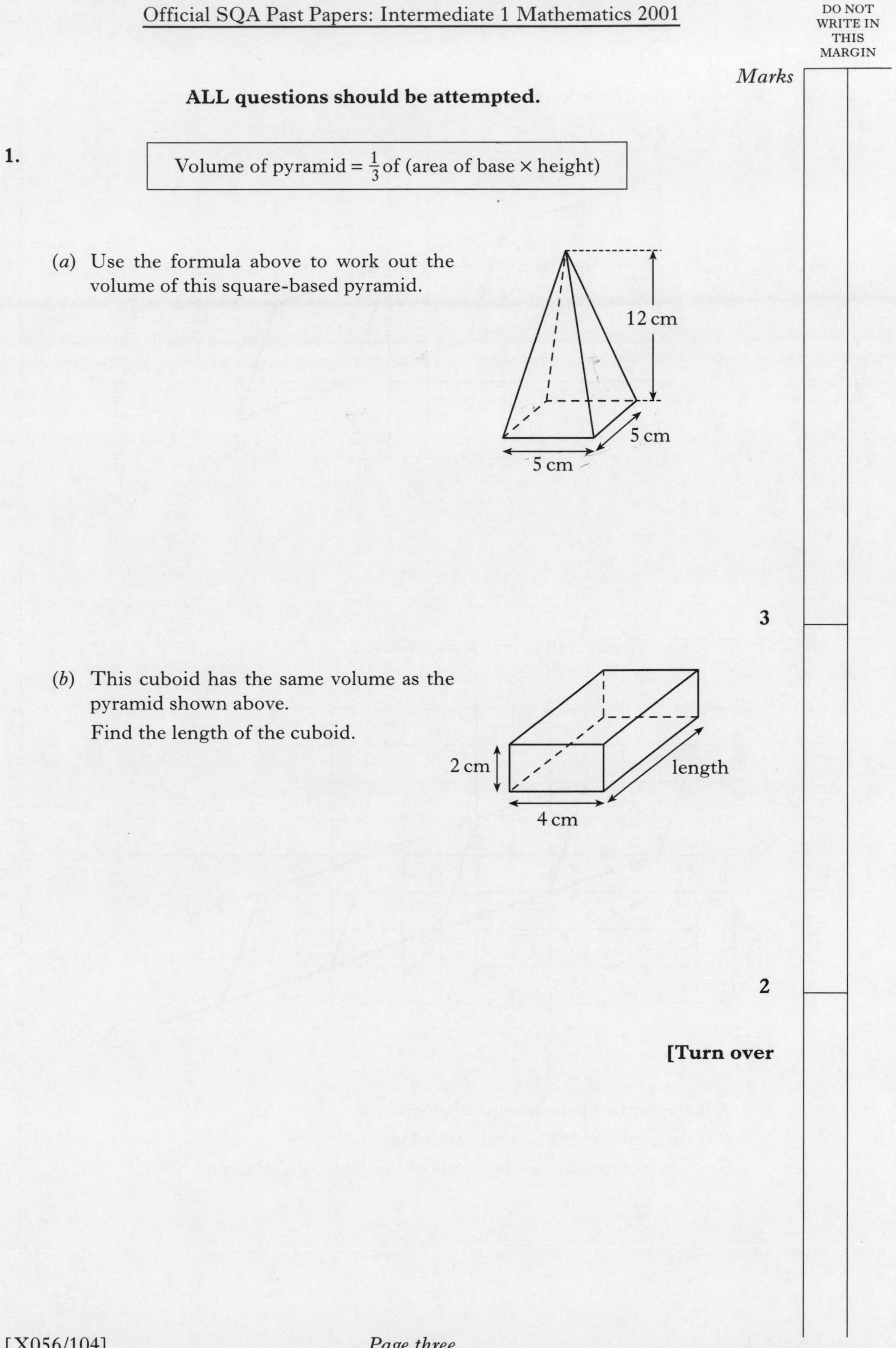

ALL questions should be attempted.

1.

Volume of pyramid = $\frac{1}{3}$ of (area of base × height)

(*a*) Use the formula above to work out the volume of this square-based pyramid. **3**

(*b*) This cuboid has the same volume as the pyramid shown above.

Find the length of the cuboid. **2**

[Turn over

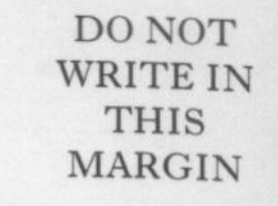

Marks

2. (*a*) Write down the coordinates of the point A marked on this diagram.

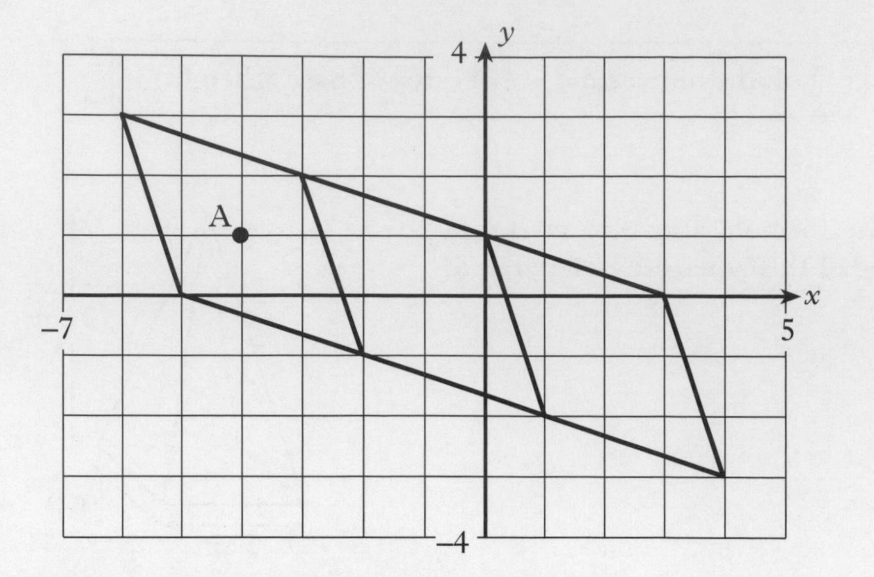

1

(*b*) The pattern of parallelograms continues.

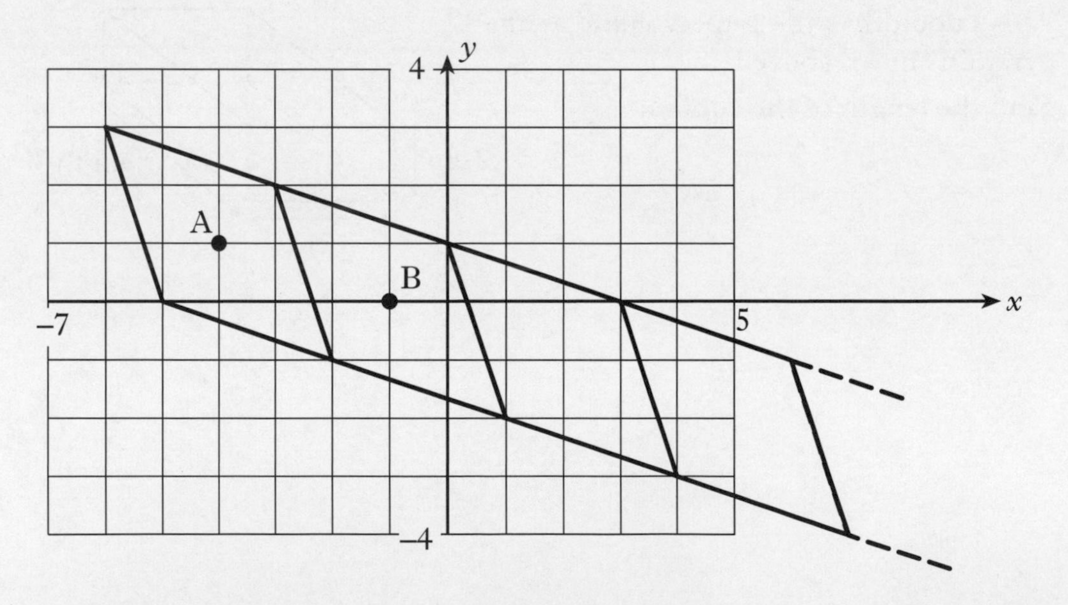

A is the centre of the first parallelogram.

B is the centre of the second parallelogram.

Find the coordinates of the centre of the sixth parallelogram.

2

DO NOT WRITE IN THIS MARGIN

Marks

3. A group of swimmers record

- the number of lengths they swim in each training session
- their personal best time (in seconds) for swimming 100 metres in competition.

The scattergraph shows the results.

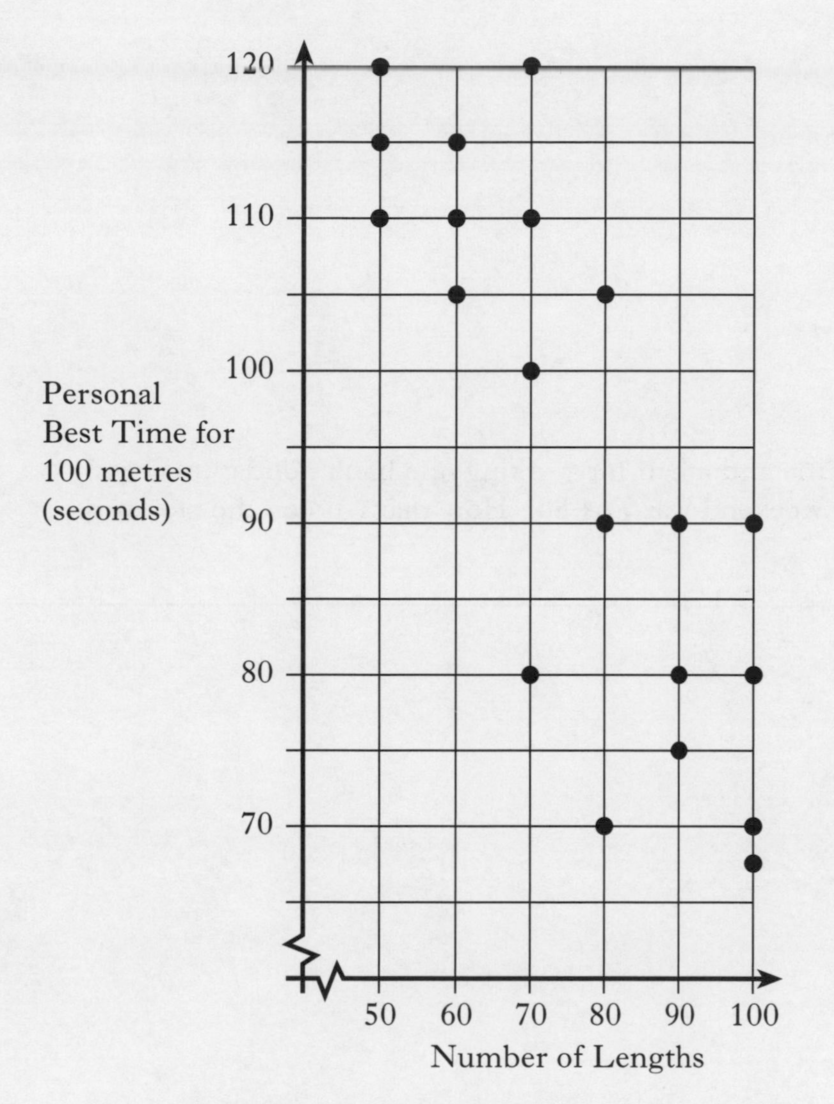

(*a*) Draw a line of best fit through the points on the graph. **1**

(*b*) Use the graph to estimate the personal best time of a swimmer who swims 75 lengths in each training session. **1**

[Turn over

DO NOT WRITE IN THIS MARGIN

Marks

4. Laura works part-time in a petrol station.

She works Monday to Friday from 5.30 pm until 10.15 pm.

Her basic rate of pay is £3·60 per hour.

(*a*) Calculate her weekly wage. **3**

(*b*) Laura was paid time and a half for working one bank holiday weekend. Her wage for the weekend was £64·80. How many hours did she work that weekend? **2**

DO NOT WRITE IN THIS MARGIN

Marks

5. Andrea leaves home in Perth at 7 am and drives 40 miles to Edinburgh Airport where she then catches a flight to Dublin. Her journey is shown on the graph below.

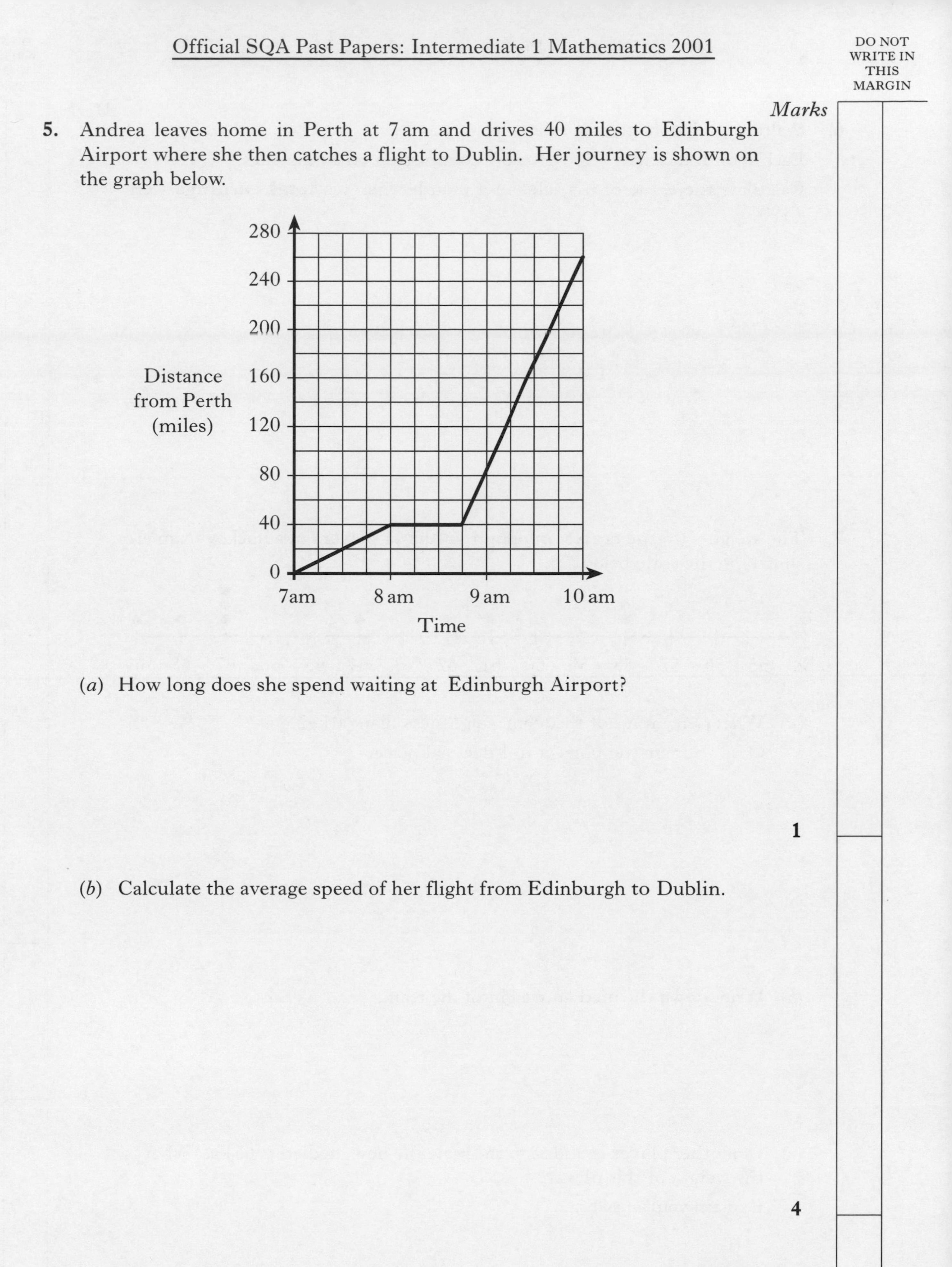

(*a*) How long does she spend waiting at Edinburgh Airport? **1**

(*b*) Calculate the average speed of her flight from Edinburgh to Dublin. **4**

[Turn over

DO NOT WRITE IN THIS MARGIN

Marks

6. Walter is a double glazing salesman.

Each month he earns £500 **plus** 5% commission on all his sales.

Calculate the value of his sales in a month when his **total earnings** were £1900.

3

7. The weights (to the nearest kilogram) of the 11 players in a hockey team are shown on the scale below.

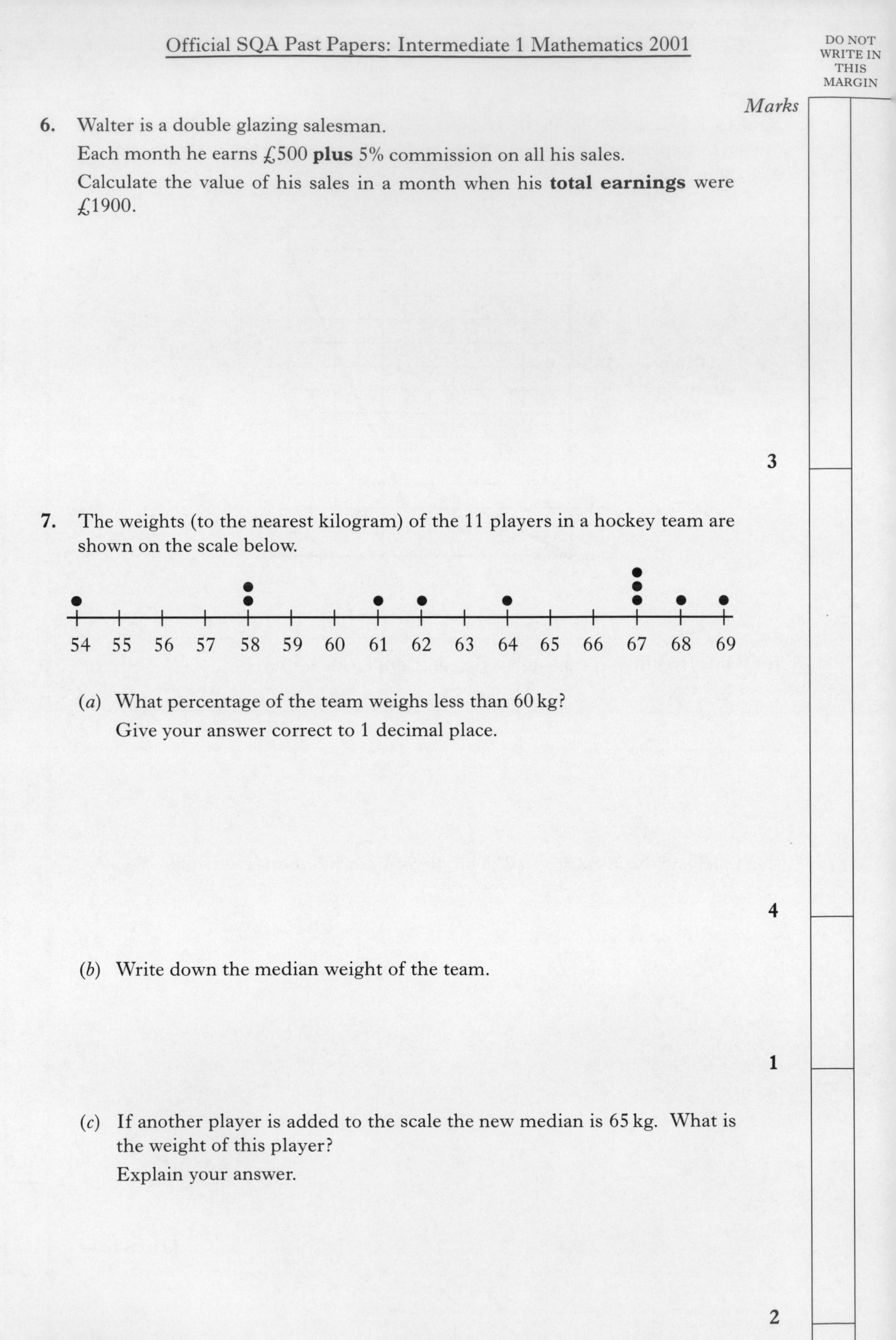

(*a*) What percentage of the team weighs less than 60 kg?

Give your answer correct to 1 decimal place.

4

(*b*) Write down the median weight of the team.

1

(*c*) If another player is added to the scale the new median is 65 kg. What is the weight of this player?

Explain your answer.

2

DO NOT WRITE IN THIS MARGIN

Marks

8. The box office takings at cinemas in the UK and the USA from showing "The Spartans" are shown below.

"THE SPARTANS" Box Office Takings	
UK	£10 230 000
USA	$15 800 000

Exchange Rate: £1 = $1·52

Change the box office takings in the USA to pounds sterling.

Give your answer to the nearest thousand pounds.

3

9. A bypass is being built to reduce the traffic passing through Steevley as shown in the diagram.

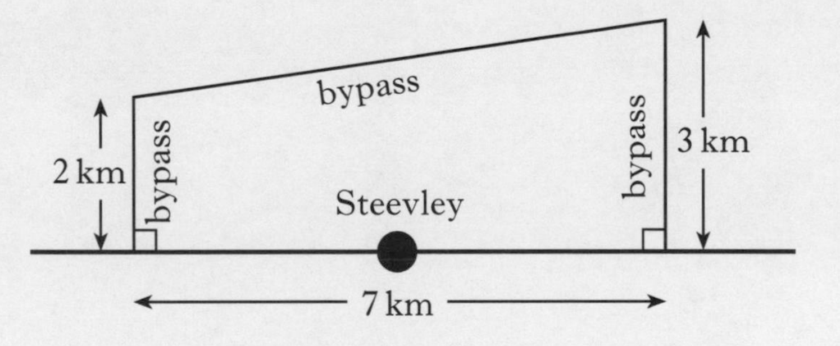

Calculate the total length of the bypass.

Do not use a scale drawing.

4

DO NOT WRITE IN THIS MARGIN

Marks

10. This sign is in the shape of a rectangle and a semi-circle.

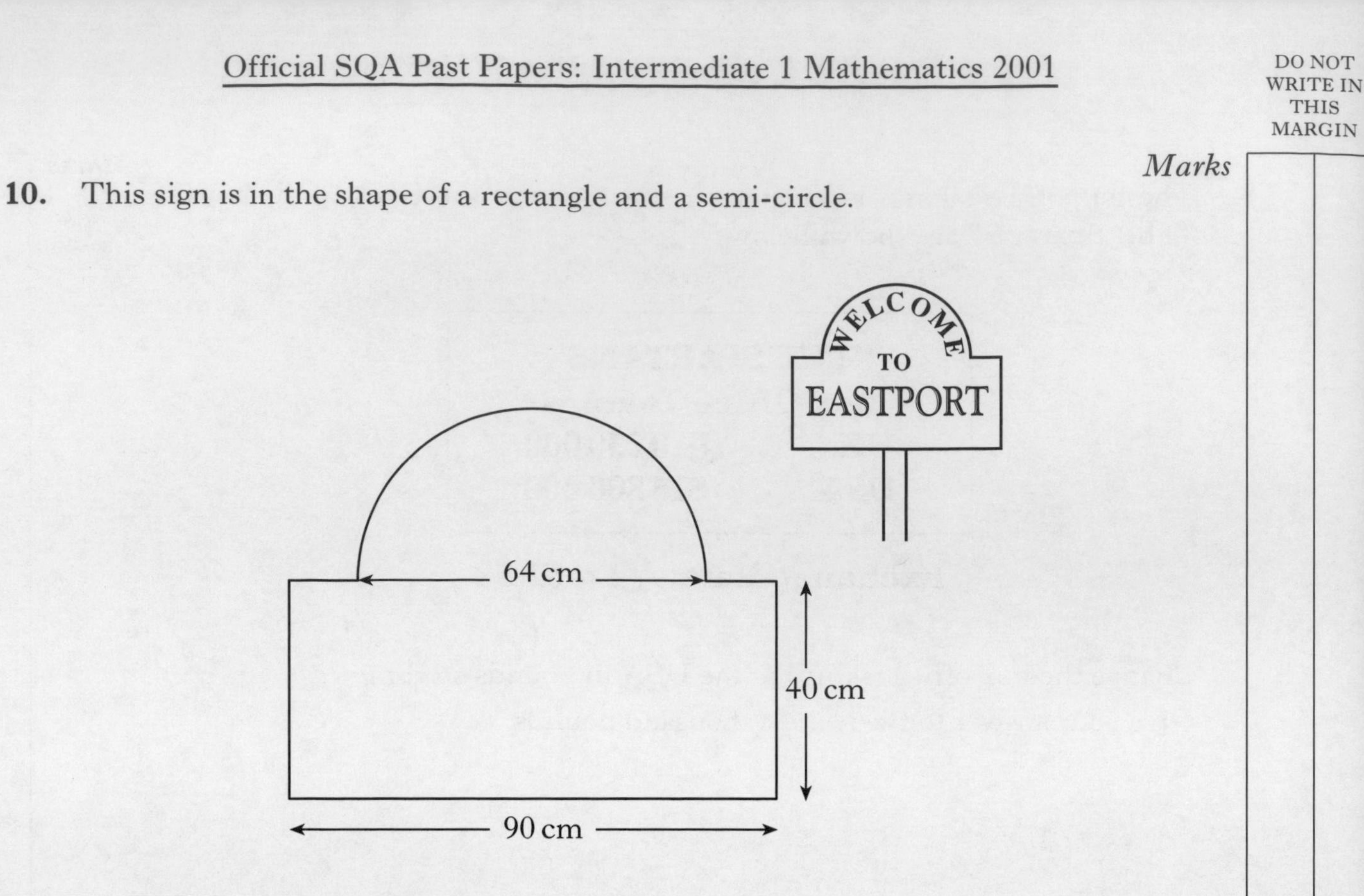

Calculate the area of the sign.

Give your answer to the nearest square centimetre.

5

DO NOT WRITE IN THIS MARGIN

Marks

11. During a 12 day period in June, Frank records the amount of rainfall (in millimetres) each day.

The rainfall figures are shown below.

1, 8, 3, 9, 11, 8, 2, 4, 9, 11, 12, 7

Calculate the interquartile range of the rainfall figures. **4**

[*END OF QUESTION PAPER*]

DO NOT WRITE IN THIS MARGIN

ADDITIONAL SPACE FOR ANSWERS

[BLANK PAGE]

FOR OFFICIAL USE

Total mark

X101/102

NATIONAL QUALIFICATIONS 2002

MONDAY, 27 MAY
1.00 PM – 1.35 PM

MATHEMATICS
INTERMEDIATE 1
Units 1, 2 and Applications of Mathematics
Paper 1
(Non-calculator)

Fill in these boxes and read what is printed below.

Full name of centre

Town

Forename(s)

Surname

Date of birth
Day Month Year

Scottish candidate number

Number of seat

1 **You may NOT use a calculator.**

2 Write your working and answers in the spaces provided. Additional space is provided at the end of this question-answer book for use if required. If you use this space, write clearly the number of the question involved.

3 Full credit will be given only where the solution contains appropriate working.

4 Before leaving the examination room you must give this book to the invigilator. If you do not you may lose all the marks for this paper.

FORMULAE LIST

Circumference of a circle: $C = \pi d$
Area of a circle: $A = \pi r^2$
Curved surface area of a cylinder: $A = 2\pi rh$

Theorem of Pythagoras:

c, b, a

$a^2 + b^2 = c^2$

DO NOT WRITE IN THIS MARGIN

Marks

ALL questions should be attempted.

1. (*a*) Find $5{\cdot}22 \div 9$. **1**

(*b*) Find $\frac{2}{5}$ of £80. **1**

2. Find the volume of this cuboid.

APPLE JUICE

12 cm

6 cm

7 cm

2

[Turn over

DO NOT WRITE IN THIS MARGIN

Marks

3. The graph shows the amount of full cream and semi-skimmed milk sold by a supermarket from 1990 to 2001.

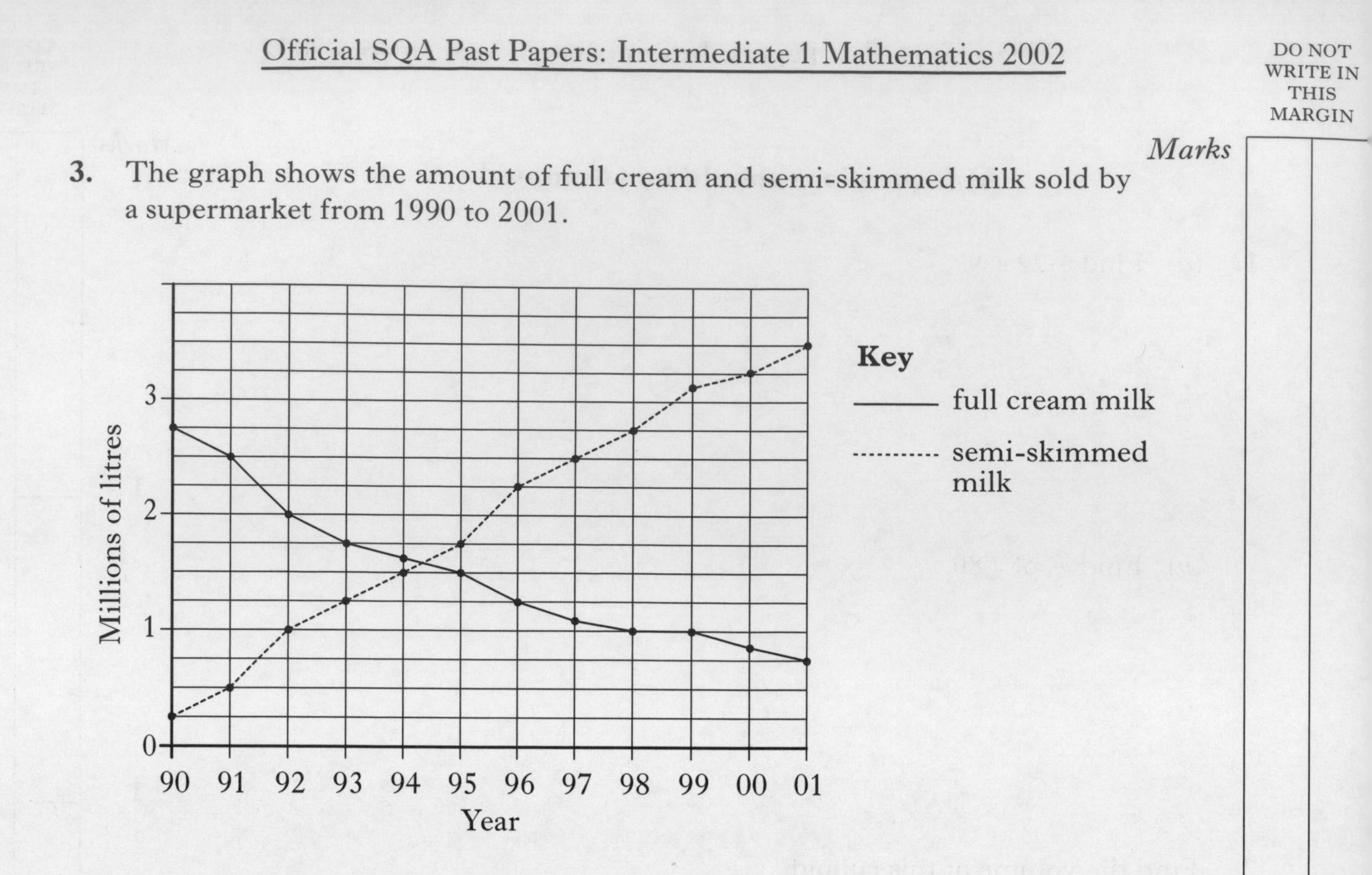

(*a*) How much semi-skimmed milk was sold in 1991? **1**

(*b*) Describe the trend in sales of **both** kinds of milk. **1**

DO NOT WRITE IN THIS MARGIN

Marks

4. This information appears on a box of chocolates.

Nutritional Information

per 100 grams

Energy	489 kJ
Protein	6·28 g
Carbohydrate	57·1 g
Fat	25·6 g

How much fat is in 300 grams of the chocolates?

2

[Turn over

DO NOT WRITE IN THIS MARGIN

Marks

5. Geeta is buying a new car. Her local garage has the following special offer on new cars.

SPECIAL OFFER on new cars 3 ITEMS FREE

Choose any THREE of these items up to a maximum value of £850

CD player	*£150*
Air Conditioning	*£300*
One year's Insurance	*£400*
Central Locking	*£200*
Electric Sunroof	*£350*

(*a*) One combination of items is shown in the table below.

CD player	*Air Conditioning*	*One year's Insurance*	*Central Locking*	*Electric Sunroof*	*Total Value*
✓		✓	✓		£750

Complete the table to show **all** the possible combinations of items available under this special offer. **3**

(*b*) Geeta wants all five of these items.
She is willing to pay for the extra two items.
What is the least amount she must pay? **2**

DO NOT WRITE IN THIS MARGIN

Marks

6. Sam works in a shop. He is paid £5 an hour. He works 36 hours a week. Complete his payslip below.

Name: Sam MacDonald		**Week ending:** 18/5/02
Hours worked 36	**Hourly rate** £5·00	**Gross pay**
Tax £16·91	**National Insurance** £10·40	**Total Deductions**
		Net Pay

3

[Turn over

DO NOT WRITE IN THIS MARGIN

Marks

7. The scale drawing shows the positions of Aberdeen and the oilrig Nordic Bravo.

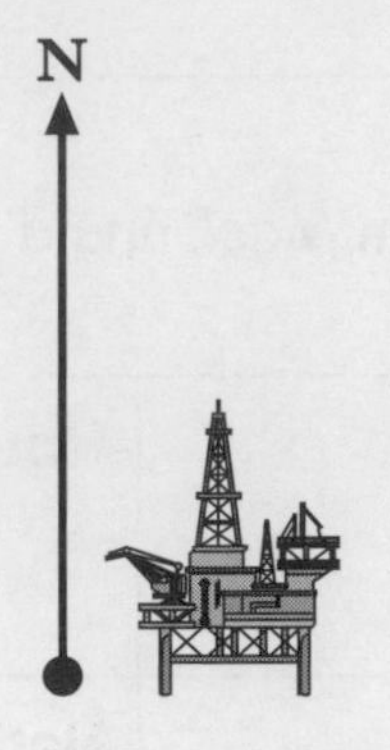

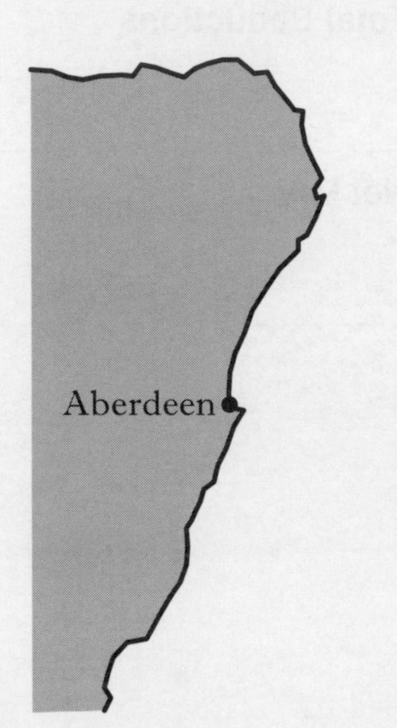

Scale: 1 cm to 30 km

Use the scale drawing to find the distance and bearing of Aberdeen **from Nordic Bravo**.

4

DO NOT WRITE IN THIS MARGIN

Marks

8. The full premium for John to insure his car last year was £480.

This year the premium has increased by one third.

John also receives a 20% discount on **this year's** premium.

How much will John pay to insure his car this year?

4

[Turn over

Marks

9. The attendances at six football matches are listed below.

7000 10 000 64 000 11 000 10 000 12 000

(*a*) Find the mean attendance. **2**

(*b*) Find the median attendance. **2**

(*c*) Which of the averages gives a truer picture of the above attendances — the mean or the median?

Give a reason for your answer. **1**

DO NOT WRITE IN THIS MARGIN

Marks

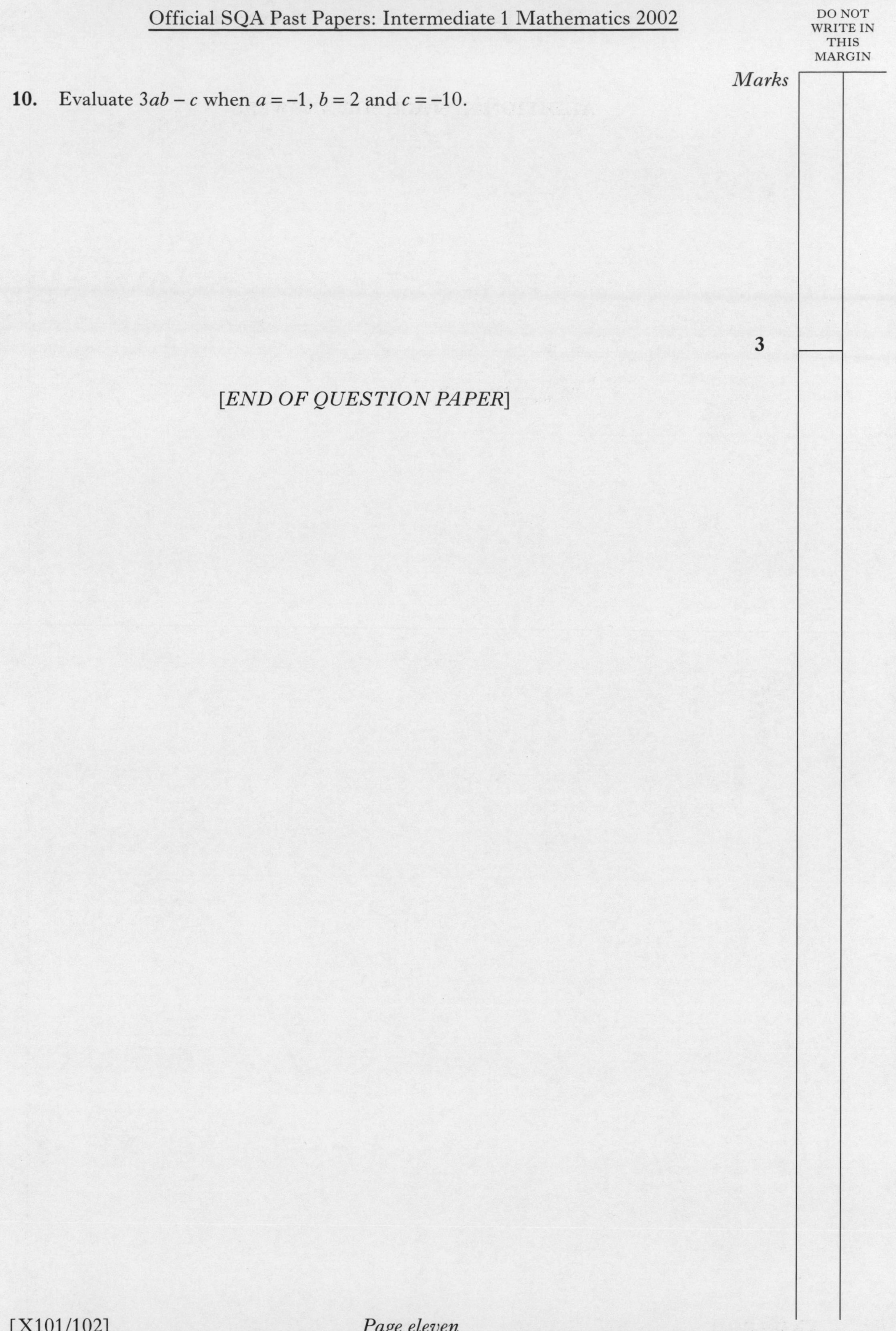

10. Evaluate $3ab - c$ when $a = -1$, $b = 2$ and $c = -10$. **3**

[*END OF QUESTION PAPER*]

ADDITIONAL SPACE FOR ANSWERS

FOR OFFICIAL USE

Total marks

X101/104

NATIONAL QUALIFICATIONS 2002

MONDAY, 27 MAY
1.55 PM – 2.50 PM

MATHEMATICS
INTERMEDIATE 1
Units 1, 2 and Applications of Mathematics Paper 2

Fill in these boxes and read what is printed below.

Full name of centre

Town

Forename(s)

Surname

Date of birth
Day Month Year

Scottish candidate number

Number of seat

1 **You may use a calculator.**

2 Write your working and answers in the spaces provided. Additional space is provided at the end of this question-answer book for use if required. If you use this space, write clearly the number of the question involved.

3 Full credit will be given only where the solution contains appropriate working.

4 Before leaving the examination room you must give this book to the invigilator. If you do not you may lose all the marks for this paper.

FORMULAE LIST

Circumference of a circle: $C = \pi d$

Area of a circle: $A = \pi r^2$

Curved surface area of a cylinder: $A = 2\pi rh$

Theorem of Pythagoras:

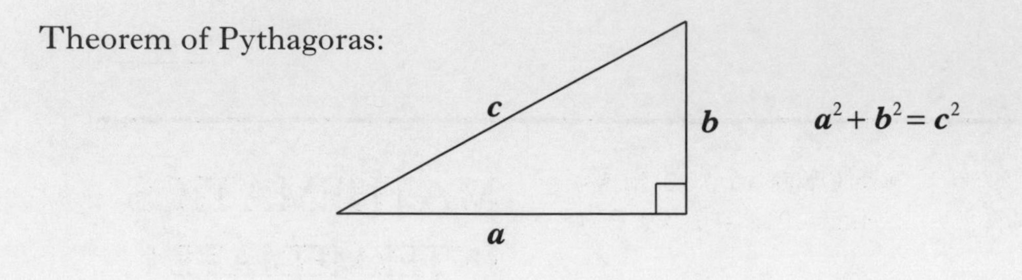

$$a^2 + b^2 = c^2$$

DO NOT WRITE IN THIS MARGIN

Marks

ALL questions should be attempted.

1. A letter is chosen at random from the letters of the word

MATHEMATICS.

What is the probability that the chosen letter is **M**? **1**

2. The table below shows the **monthly repayments** to be made when money is borrowed from the Inverness Building Society.

Loan \ *Term*	*Monthly Repayments*		
	20 years	*25 years*	*30 years*
£20 000	£158·64	£146·39	£136·98
£30 000	£239·26	£217·70	£206·12
£40 000	£320·03	£290·89	£273·73

Katy needs to borrow **£30 000** to buy a flat.

(*a*) She decides to repay the loan over **30 years**.

(i) How much will she pay each month? **1**

(ii) How much will she pay altogether? **2**

(*b*) How much would Katy have saved if she had borrowed the £30 000 over **20 years**? **2**

DO NOT WRITE IN THIS MARGIN

Marks

3. The number of copies of "The Anglers Weekly" magazine sold by a newsagent was recorded for 16 weeks.

25	23	19	22	18	45	38	23
32	25	51	27	23	30	28	42

(*a*) Complete this stem and leaf diagram using the data above.

1|8 represents 18 magazines

2

(*b*) Find the mode for this data set.

1

DO NOT WRITE IN THIS MARGIN

Marks

4. Jane is going to Switzerland and wants to change £500 into Swiss francs. Two travel agents offer the following exchange rates.

TRAVELSUN	**SOLLAIR**
£1 = 2·46 Swiss francs	**£1 = 2·50 Swiss francs**
No commission	**2% commission payable**

(*a*) How many Swiss francs would Jane receive from Travelsun for £500? **1**

(*b*) Which travel agent will give Jane more Swiss francs for her £500?
Show clearly all your working. **4**

[Turn over

DO NOT WRITE IN THIS MARGIN

Marks

5. Janis works in a shirt factory.

This flowchart is used to work out the weekly bonus she is paid for making shirts.

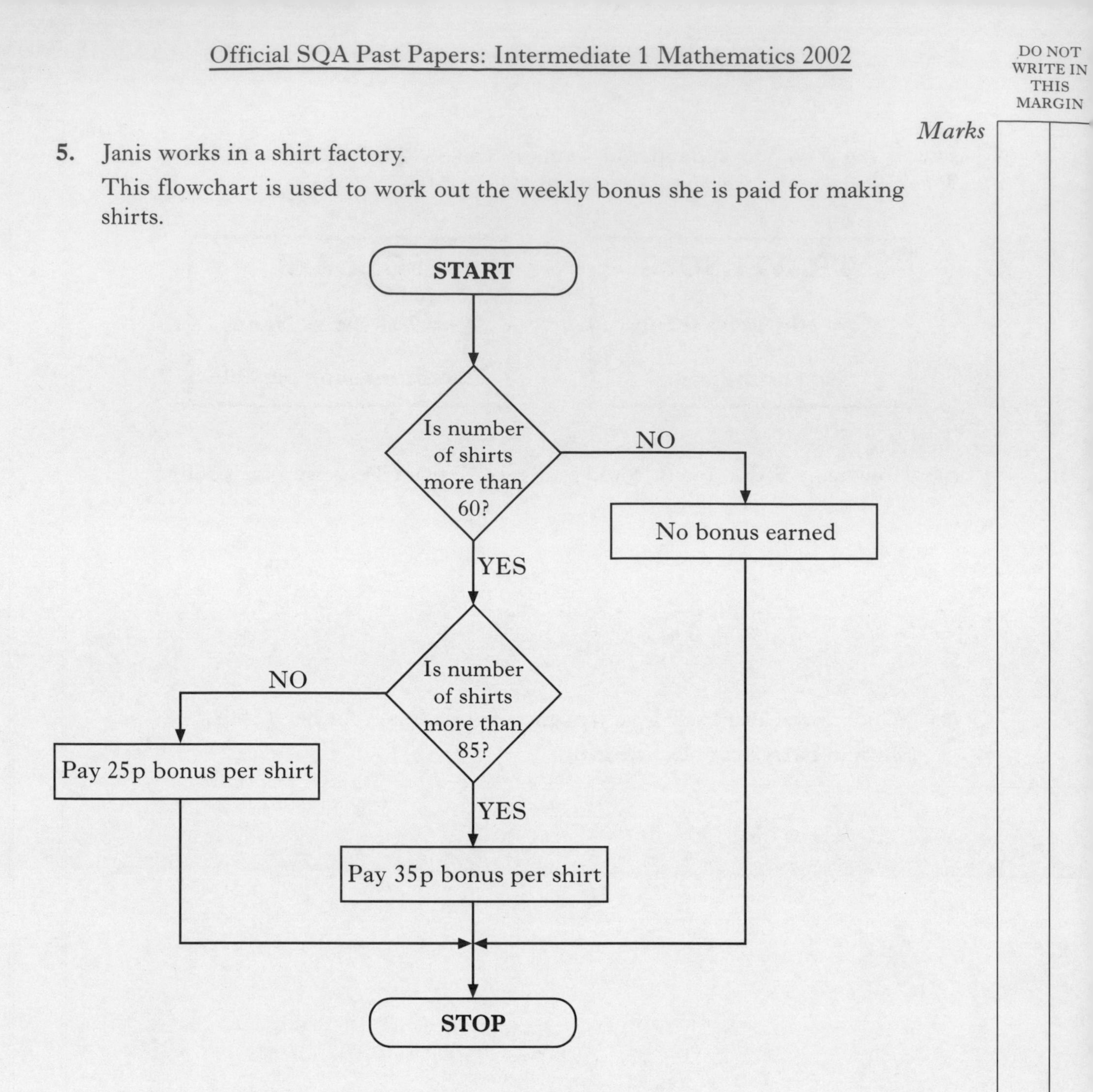

(*a*) Janis made 102 shirts one week. Calculate her bonus. **2**

(*b*) Is it possible for Janis to earn a bonus of **exactly** £25 for making shirts?

Explain your answer. **3**

DO NOT WRITE IN THIS MARGIN

Marks

6. Ali drove overnight 406 miles from Galashiels to Portsmouth to catch a ferry to France.

His average speed for the journey was 56 miles per hour.

He arrived in Portsmouth at 0630.

At what time did he leave Galashiels?

4

7. A group of students was asked how many times they had visited a cinema during the last month.

The results are shown in this frequency table.

Number of visits	*Frequency*	*Visits × Frequency*
0	104	
1	56	
2	44	
3	20	
4	10	
5	1	
	Total = 235	Total =

Complete the table above and find the mean number of visits.

Give your answer correct to 1 decimal place.

3

DO NOT WRITE IN THIS MARGIN

Marks

8. The circle shown below has centre (0,0).

The point (6,8) lies on the circle.

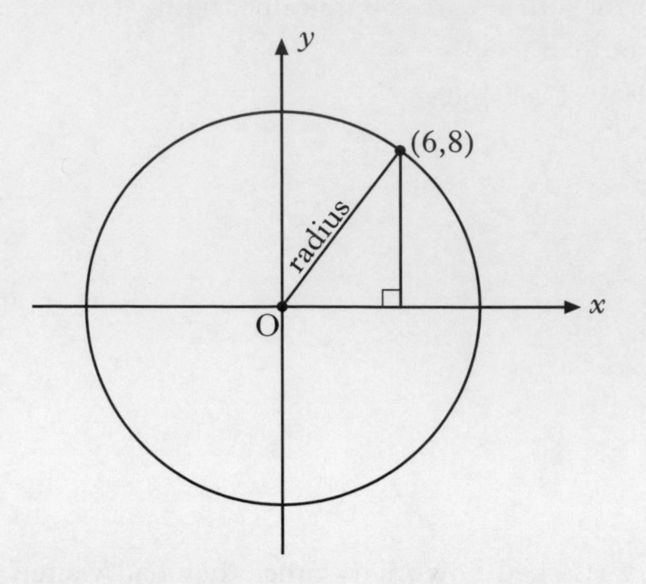

Work out the area of this circle.

4

DO NOT WRITE IN THIS MARGIN

Marks

9. Ian measures the heights of some plants which have been growing for one month.

Their heights (in millimetres) are shown below.

29 25 37 35 35 37 41 46 39 47 56 61 49

Complete the boxplot, drawn below, to show the heights of the plants.

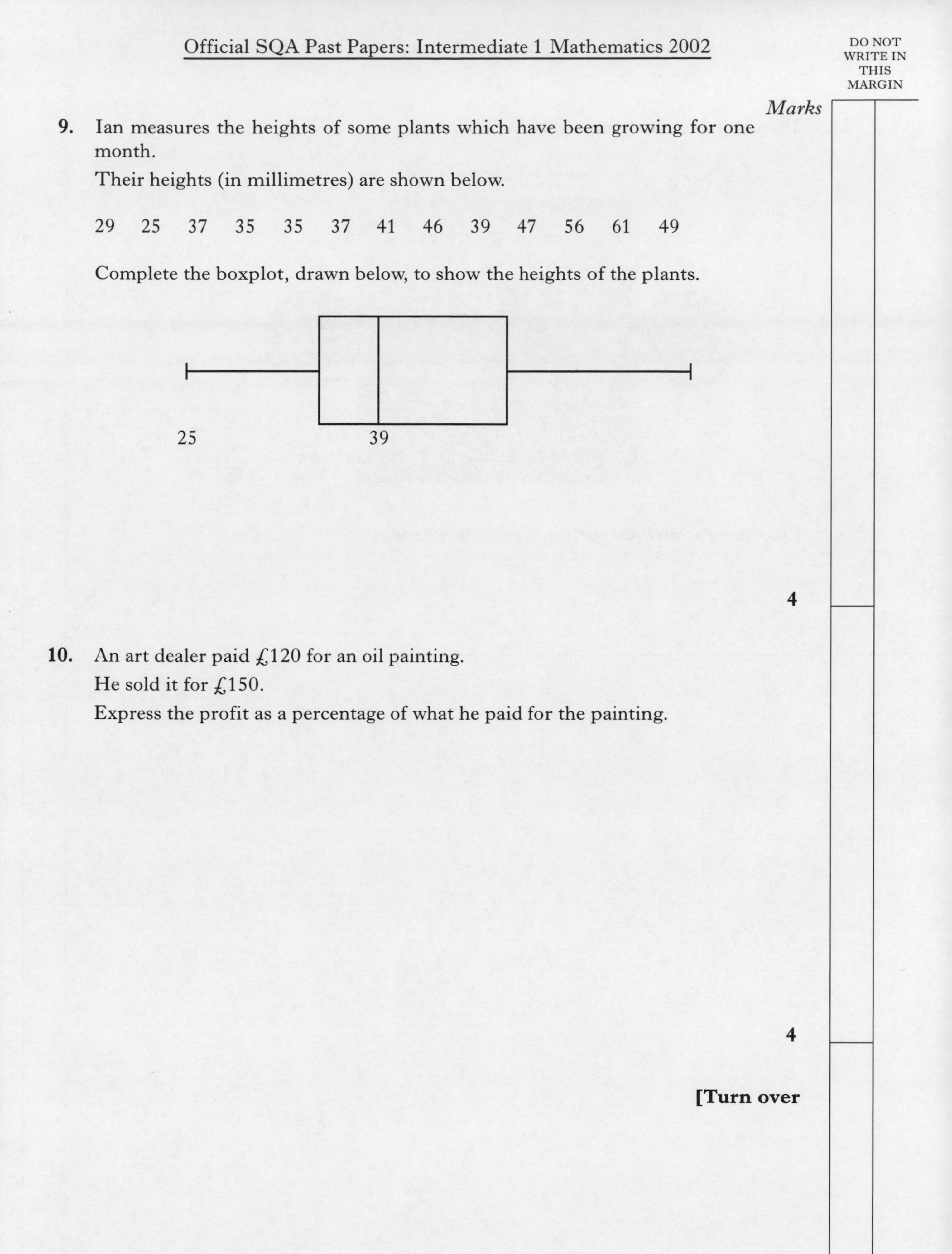

4

10. An art dealer paid £120 for an oil painting.

He sold it for £150.

Express the profit as a percentage of what he paid for the painting.

4

[Turn over

DO NOT WRITE IN THIS MARGIN

Marks

11. The diagram below shows the net of a cylinder.

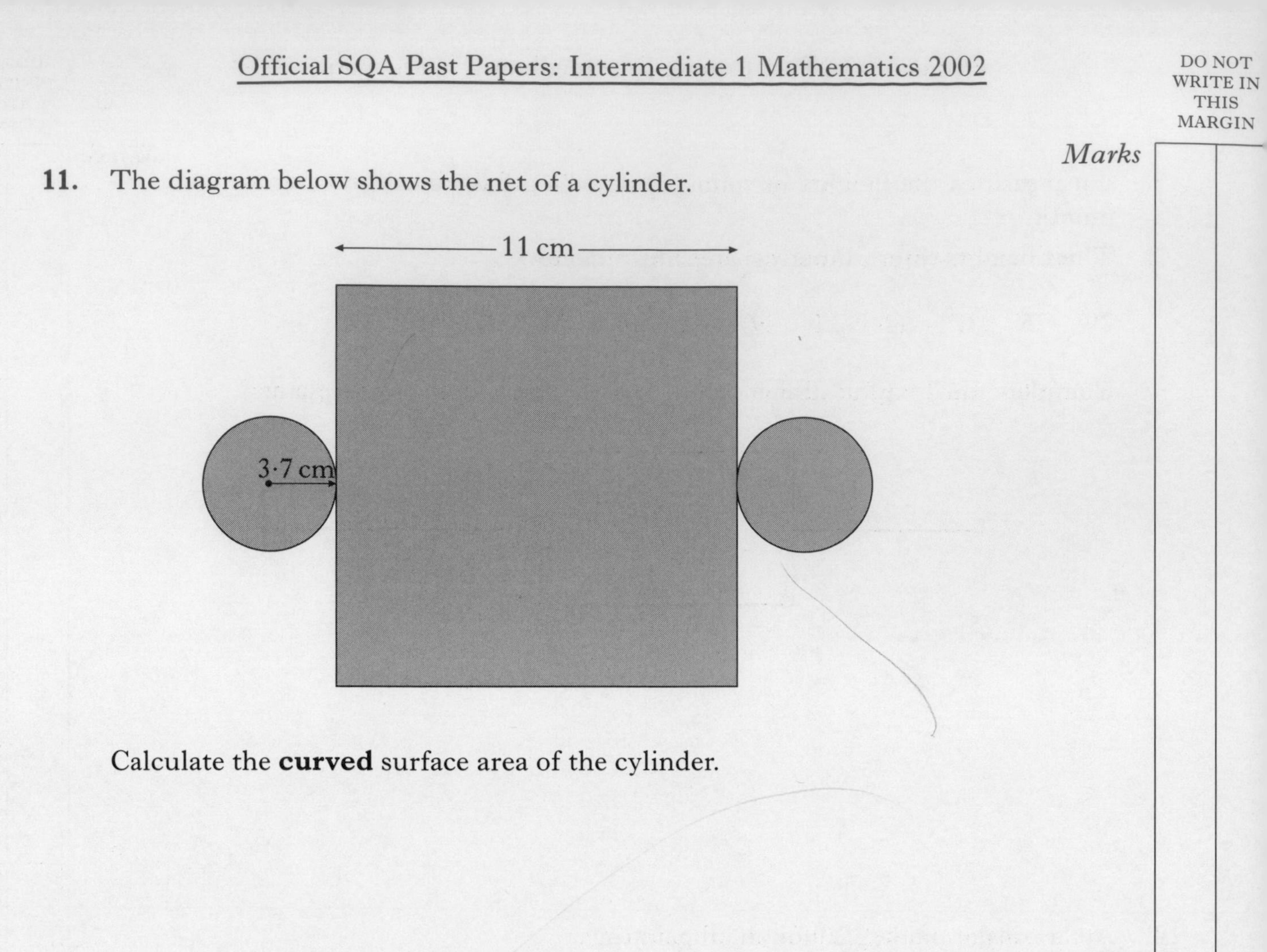

Calculate the **curved** surface area of the cylinder.

3

DO NOT WRITE IN THIS MARGIN

Marks

12. This window blind is in the shape of a rectangle with four equal semi-circles at the bottom.

It has braid down the two sides and round the bottom.

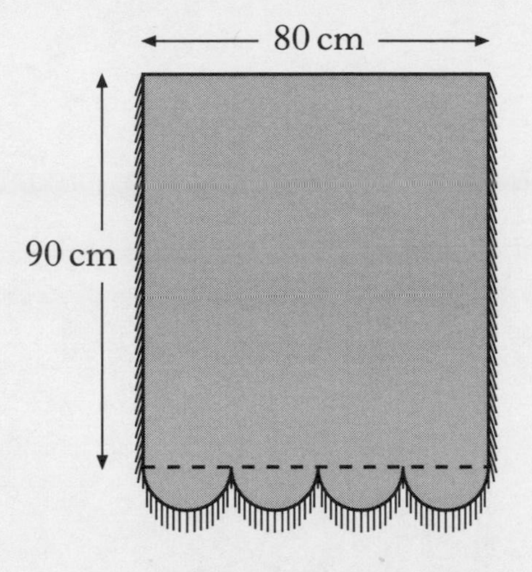

Calculate the total length of braid needed for this blind.

Give your answer to the nearest centimetre. **5**

[*END OF QUESTION PAPER*]

DO NOT WRITE IN THIS MARGIN

ADDITIONAL SPACE FOR ANSWERS

[BLANK PAGE]

FOR OFFICIAL USE

Total mark

X101/102

NATIONAL QUALIFICATIONS 2003

WEDNESDAY, 21 MAY
1.30 PM – 2.05 PM

MATHEMATICS INTERMEDIATE 1
Units 1, 2 and Applications of Mathematics
Paper 1
(Non-calculator)

Fill in these boxes and read what is printed below.

Full name of centre

Town

Forename(s)

Surname

Date of birth
Day Month Year

Scottish candidate number

Number of seat

1. **You may NOT use a calculator.**
2. Write your working and answers in the spaces provided. Additional space is provided at the end of this question-answer book for use if required. If you use this space, write clearly the number of the question involved.
3. Full credit will be given only where the solution contains appropriate working.
4. Before leaving the examination room you must give this book to the invigilator. If you do not you may lose all the marks for this paper.

LIB X101/102 6/3/3770

FORMULAE LIST

Circumference of a circle:	$C = \pi d$
Area of a circle:	$A = \pi r^2$
Curved surface area of a cylinder:	$A = 2\pi rh$

Theorem of Pythagoras:

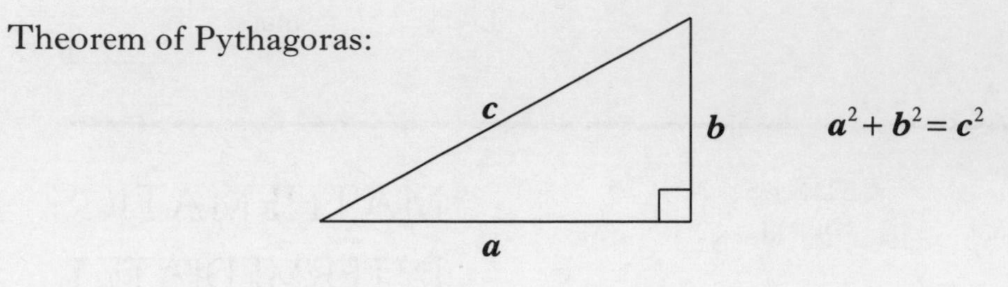

$$a^2 + b^2 = c^2$$

DO NOT WRITE IN THIS MARGIN

Marks

ALL questions should be attempted.

1. (*a*) Find $6{\cdot}23 - 3{\cdot}7$.

1

(*b*) Find 5% of £140.

1

(*c*) Find $-40 + 15$.

1

2. A rule used to calculate the cost in pounds of electricity is:

Cost = 19 + (number of units used × 0·07)

Find the cost of 600 units of electricity.

2

[Turn over

DO NOT WRITE IN THIS MARGIN

Marks

3. (*a*) An inter-city coach left Aberdeen at 10.40 am and reached Inverness at 1.25 pm.

How long did the journey take?

1

(*b*) The average speed of the coach during the journey was 40 miles per hour.

Find the distance between Aberdeen and Inverness.

3

DO NOT WRITE IN THIS MARGIN

Marks

4. The table below shows the **monthly payments** to be made when money is borrowed from a bank.

Borrowers can choose to make payments with or without payment protection.

	1 year		*3 years*		*5 years*	
Amount borrowed	*Without payment protection*	*With payment protection*	*Without payment protection*	*With payment protection*	*Without payment protection*	*With payment protection*
£3000	£267	£289	£100	£111	£67	£78
£5000	£435	£471	£157	£174	£102	£118
£7000	£609	£659	£220	£244	£142	£165
£10 000	£870	£942	£314	£348	£204	£236

(*a*) Pete borrows £3000 over 3 years **without payment protection**.
State his monthly payment.

1

(*b*) Over the 3 years, how much **extra** would Pete pay **in total** for payment protection on his loan of £3000?

2

[Turn over

DO NOT WRITE IN THIS MARGIN

Marks

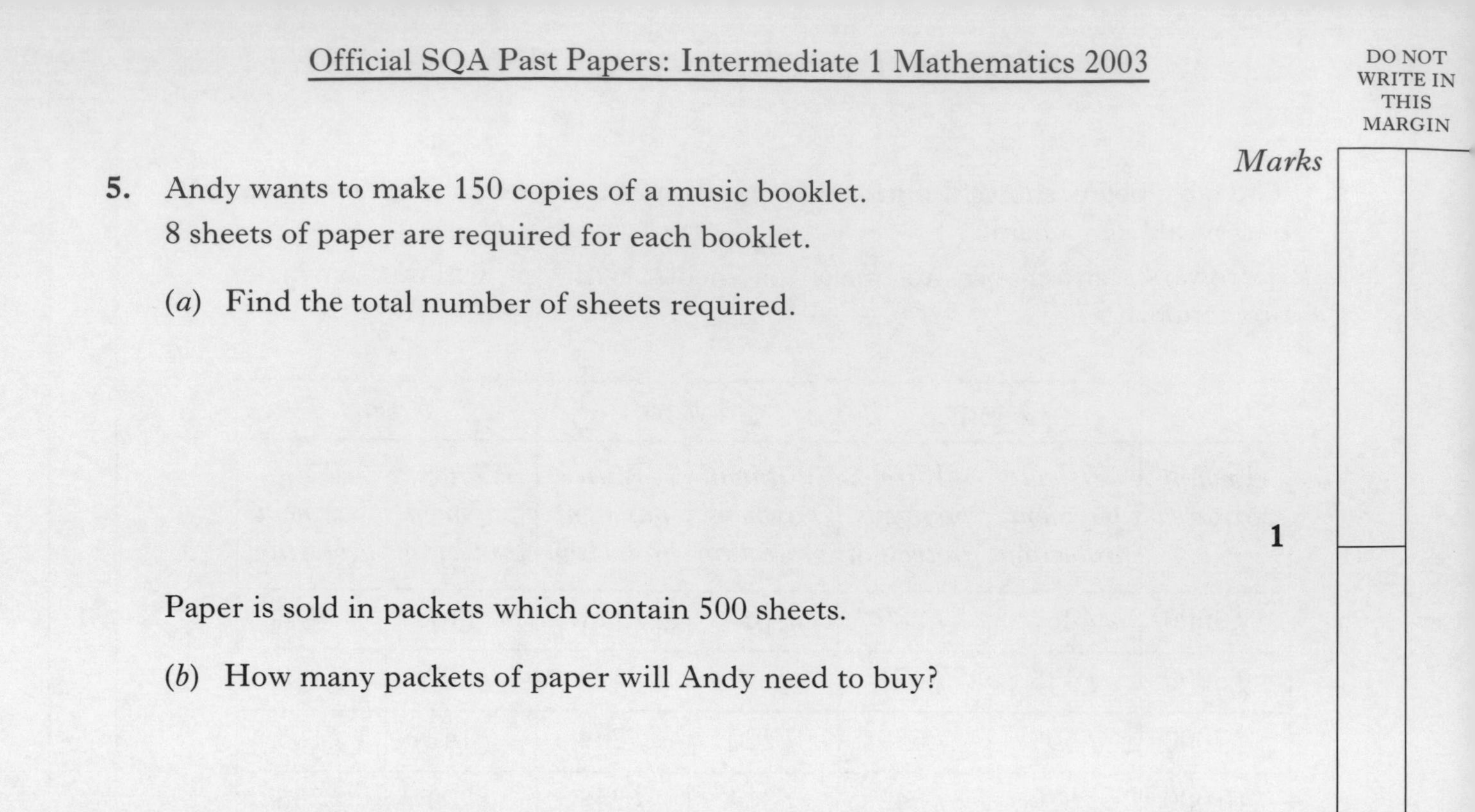

5. Andy wants to make 150 copies of a music booklet.
8 sheets of paper are required for each booklet.

(*a*) Find the total number of sheets required. **1**

Paper is sold in packets which contain 500 sheets.

(*b*) How many packets of paper will Andy need to buy? **2**

DO NOT WRITE IN THIS MARGIN

Marks

6. A spreadsheet is used to process data from a shop.

SALES ANALYSIS					
	A	B	C	D	E
1		Choc Bars	Crisps	Juice	Ice Cream
2	Monday	20	40	50	10
3	Tuesday	40	30	70	20
4	Wednesday	10	20	70	20
5	Thursday	40	50	80	10
6	Friday	60	60	80	10
7					
8					
9					

(*a*) The result of the formula =SUM(C2..C6) is to be entered in cell C8.
What would appear in cell C8?

1

(*b*) What formula would be used to enter the average daily sale of crisps in cell C9?

1

[Turn over

DO NOT WRITE IN THIS MARGIN

Marks

7. A hot air balloon is attached to the ground at A and B by two wires.

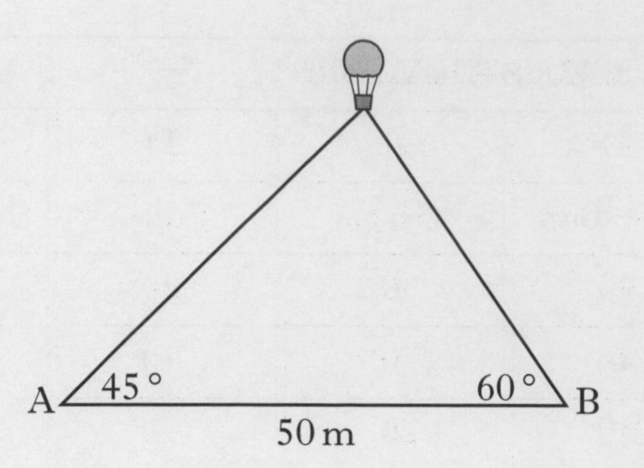

The distance from A to B is 50 metres.

The angle of elevation of the balloon is 45° from A and 60° from B.

(*a*) Make a scale drawing to show the position of the balloon.

Use a scale of 1 cm to 5 m.

2

(*b*) Use your scale drawing to find the actual height of the balloon.

2

DO NOT WRITE IN THIS MARGIN

Marks

8. In a local election the number of votes for each of the four candidates is shown in the table below.

Candidate	*Votes*
Smith	380
Patel	240
Green	100
Jones	170

On the grid below, draw a bar graph to show this information.

4

9. Five staff work in an office.

Three of them are female.

What percentage of the staff is female?

3

DO NOT WRITE IN THIS MARGIN

Marks

10. This is a multiplication square.

8	×	5	=	40
×		×		×
10	×	–2	=	–20
=		=		=
80	×	–10	=	–800

(*a*) Complete this multiplication square.

3	×	–7	=	
×		×		×
–1	×	5	=	
=		=		=
	×		=	

2

DO NOT WRITE IN THIS MARGIN

Marks

10. (continued)

(*b*) Complete this multiplication square.

–5	×		=	
×		×		×
	×		=	–12
=		=		=
	×	–8	=	–120

3

[*END OF QUESTION PAPER*]

DO NOT WRITE IN THIS MARGIN

ADDITIONAL SPACE FOR ANSWERS

FOR OFFICIAL USE

Total mark

X101/104

NATIONAL QUALIFICATIONS 2003

WEDNESDAY, 21 MAY
2.25 PM – 3.20 PM

MATHEMATICS
INTERMEDIATE 1
Units 1, 2 and Applications of Mathematics Paper 2

Fill in these boxes and read what is printed below.

Full name of centre

Town

Forename(s)

Surname

Date of birth
Day Month Year

Scottish candidate number

Number of seat

1 **You may use a calculator.**

2 Write your working and answers in the spaces provided. Additional space is provided at the end of this question-answer book for use if required. If you use this space, write clearly the number of the question involved.

3 Full credit will be given only where the solution contains appropriate working.

4 Before leaving the examination room you must give this book to the invigilator. If you do not you may lose all the marks for this paper.

LIB X101/104 6/3/3770

FORMULAE LIST

Circumference of a circle: $C = \pi d$
Area of a circle: $A = \pi r^2$
Curved surface area of a cylinder: $A = 2\pi rh$

Theorem of Pythagoras:

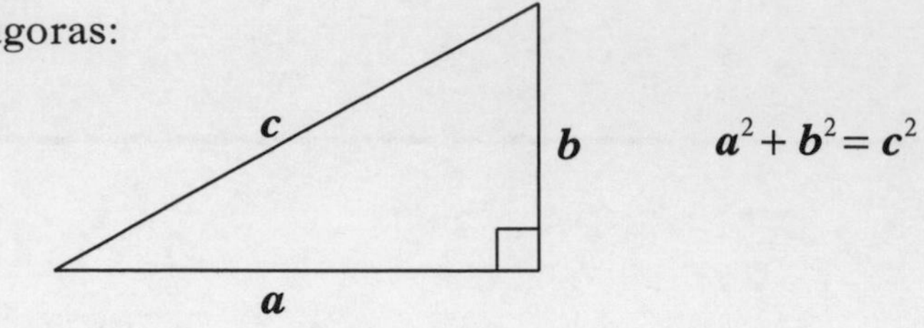

$a^2 + b^2 = c^2$

DO NOT WRITE IN THIS MARGIN

Marks

ALL questions should be attempted.

1. A day in December is chosen at random for a youth club outing.

Find the probability that a **Saturday** is chosen.

DECEMBER

Mon	Tue	Wed	Thu	Fri	Sat	Sun
1	2	3	4	5	6	7
8	9	10	11	12	13	14
15	16	17	18	19	20	21
22	23	24	25	26	27	28
29	30	31				

1

[Turn over

Marks

2. This network diagram shows the distances, in metres, between the Post Office and various buildings in a town.

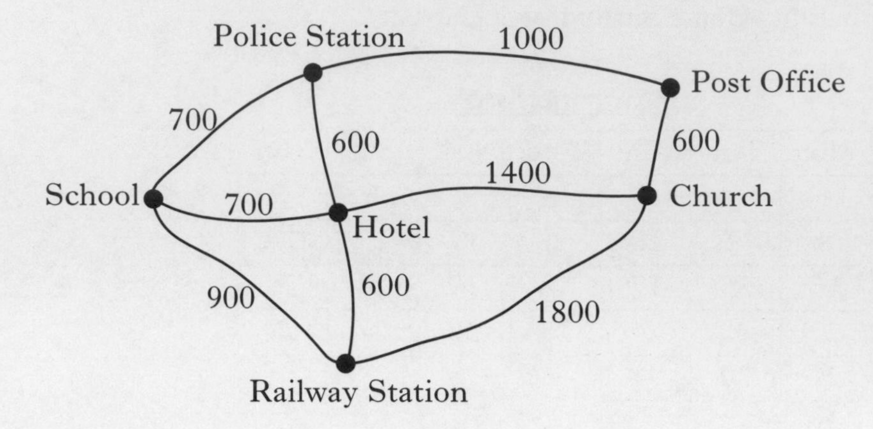

(*a*) State the **order** of the node at the Hotel. **1**

(*b*) What is the length of the shortest route from the Post Office to the Railway Station? **1**

Marks

3. Lisa is paid £7·20 per hour for a basic 35 hour week.

One week she works a total of 39 hours.

Overtime worked is paid at time and a half.

Calculate Lisa's gross pay.

4

[Turn over

DO NOT WRITE IN THIS MARGIN

Marks

4. The income of each employee in a company is shown in this frequency table.

Income £	*Frequency*	*Income* × *Frequency*
10 000	2	
12 000	3	
14 000	5	
16 000	8	
18 000	7	
	Total = 25	Total =

(*a*) Write down the modal income. **1**

(*b*) Complete the table above and find the mean income. **3**

DO NOT WRITE IN THIS MARGIN

Marks

5. A room in the Hotel Royale in Paris costs 130 euros per night.
The exchange rate is 1·58 euros to the pound.

(*a*) Find the cost of the hotel room per night in pounds and pence. **3**

Mr and Mrs McQueen are going to Paris.
Their return flights cost £59 each.

(*b*) Find the total cost of their flights and a 3 night stay at the Hotel Royale in pounds and pence. **2**

[Turn over

DO NOT WRITE IN THIS MARGIN

Marks

6. The population of Scotland in 2001 was 5 062 000.

The pie chart shows the age distribution of the population in 2001.

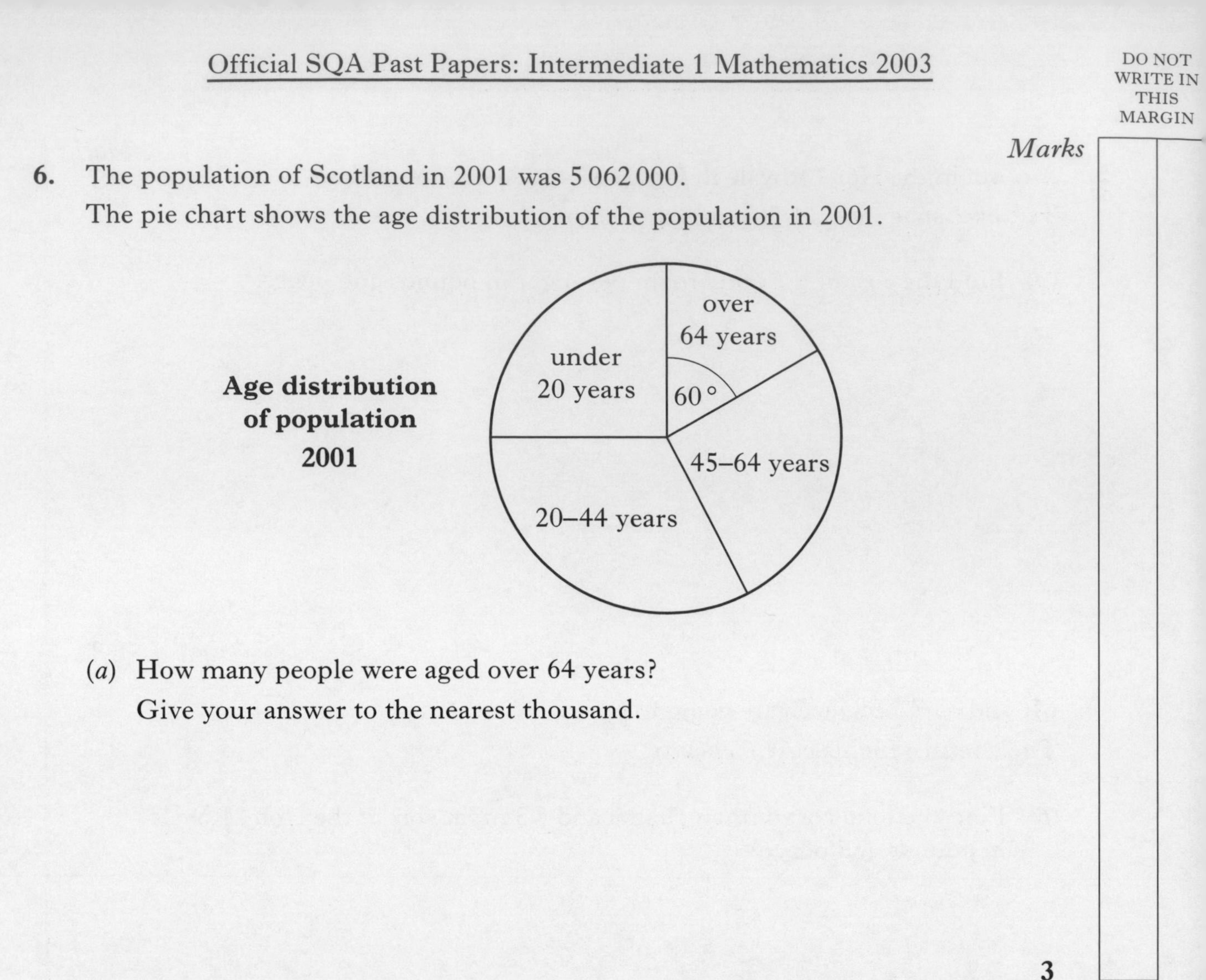

(*a*) How many people were aged over 64 years?

Give your answer to the nearest thousand. **3**

(*b*) The pie chart below shows the age distribution of the population of Scotland in 1901.

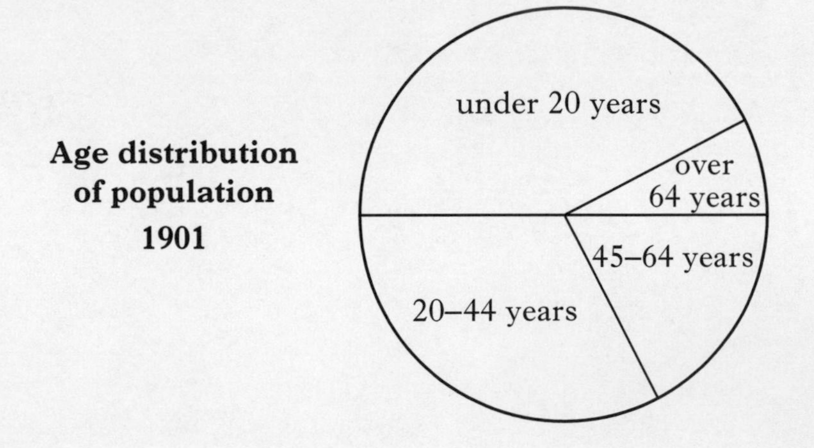

Describe the differences in the age distributions of the population of Scotland in 1901 and 2001. **2**

DO NOT WRITE IN THIS MARGIN

Marks

7. The diagram below shows two bars of soap.
Each bar is in the shape of a cuboid.

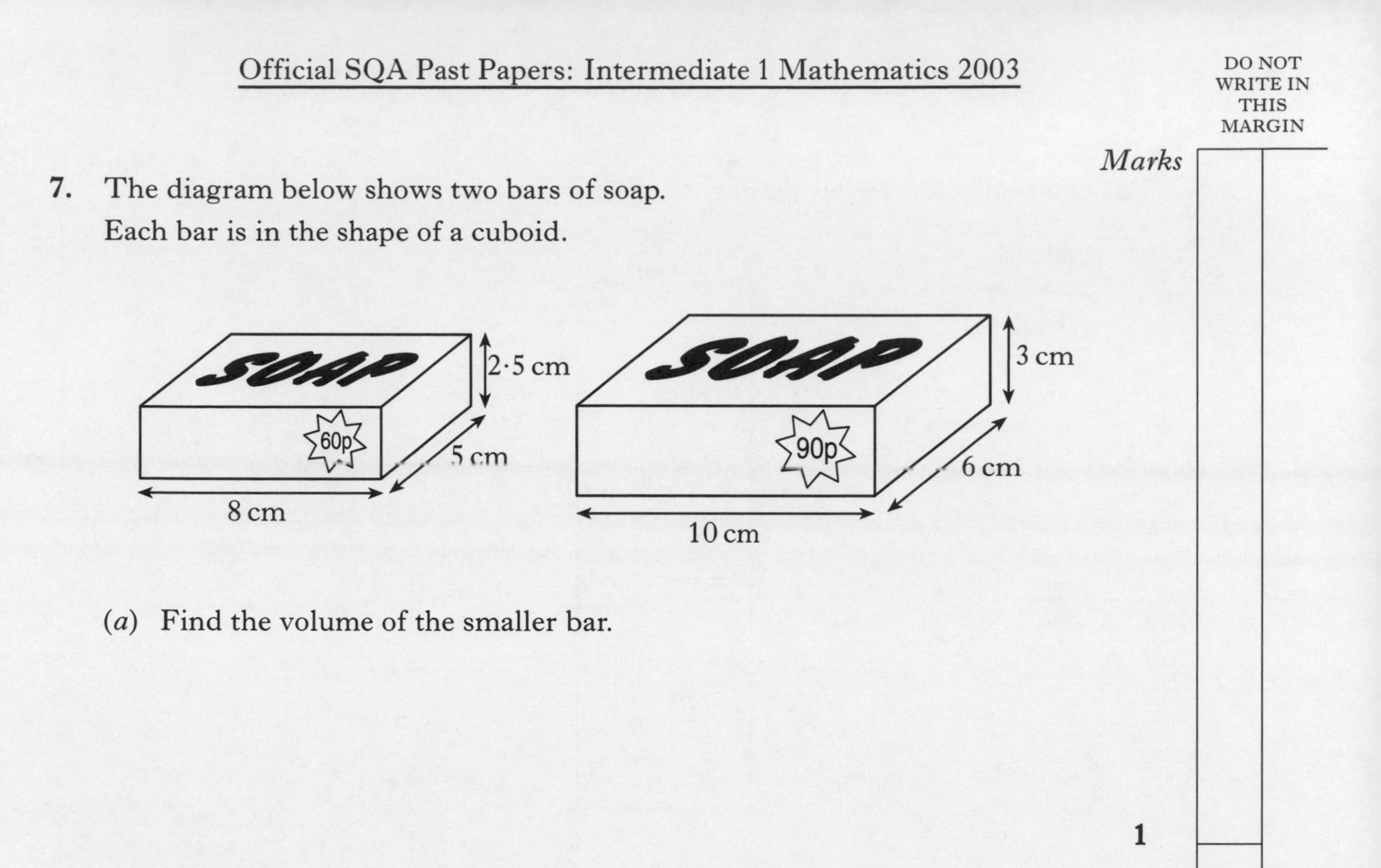

(*a*) Find the volume of the smaller bar. **1**

(*b*) The smaller bar costs 60 pence.
Find the cost per cubic centimetre of the smaller bar. **1**

(*c*) The larger bar costs 90 pence.
Which bar of soap gives better value for money?
Explain clearly the reason for your answer. **3**

[Turn over

DO NOT WRITE IN THIS MARGIN

Marks

8. The diagram below shows the net of a solid shape.

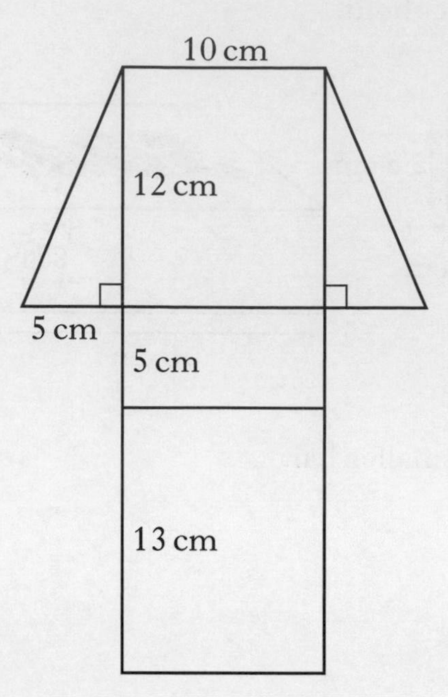

(*a*) Name the solid shape formed from this net. **1**

(*b*) Calculate the surface area of the solid shape. **3**

DO NOT WRITE IN THIS MARGIN

Marks

9. The diagram shows the front view of a garage.

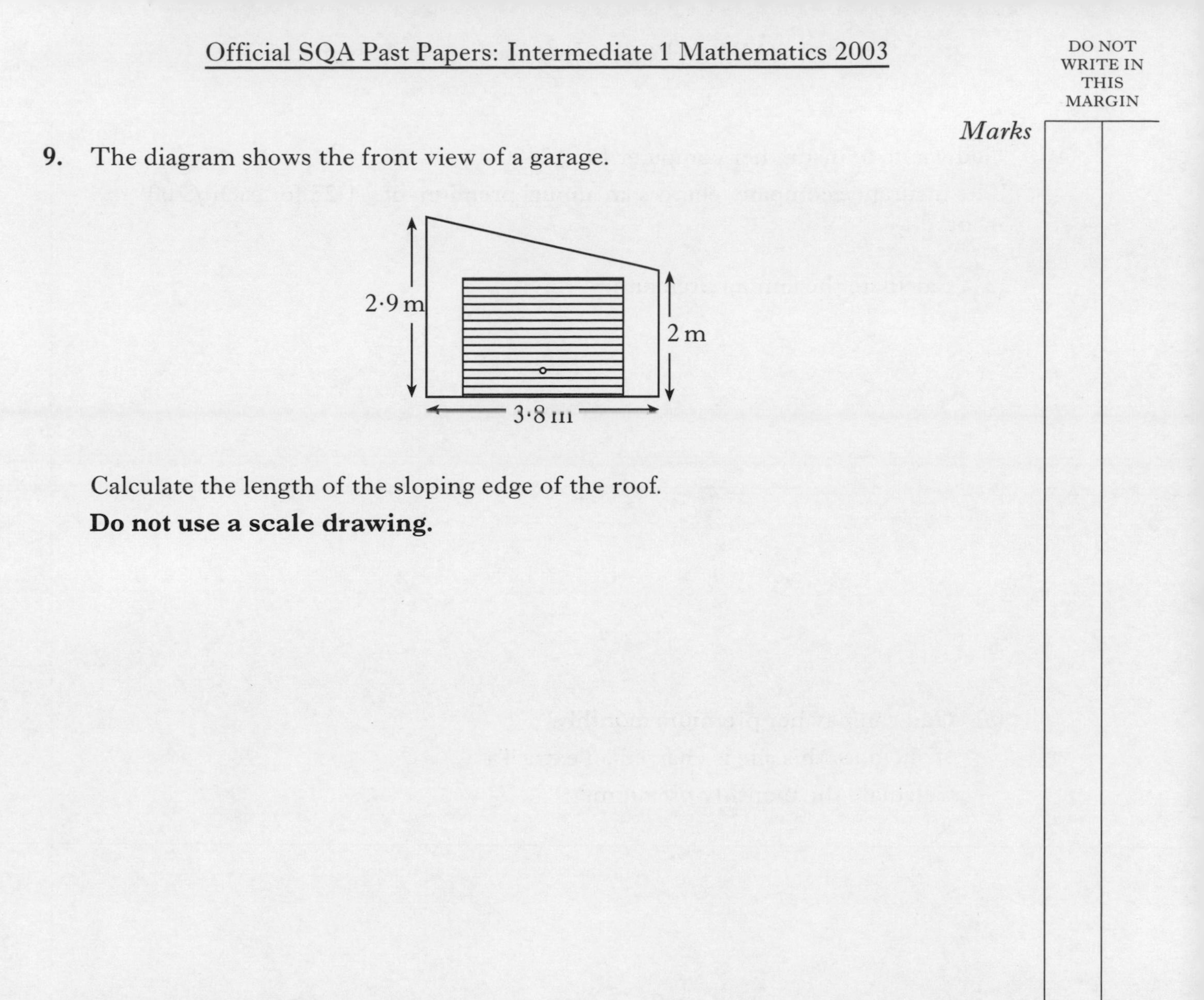

Calculate the length of the sloping edge of the roof.

Do not use a scale drawing.

3

[Turn over

DO NOT WRITE IN THIS MARGIN

Marks

10. Gail wants to insure her computer for £2400.

The insurance company charges an annual premium of £1·25 for each £100 insured.

(*a*) Calculate the annual premium. **2**

(*b*) Gail can pay her premium monthly.

If she does this she is charged an extra 4%.

Calculate the monthly premium. **3**

DO NOT WRITE IN THIS MARGIN

Marks

11. These boxplots show the sales of computer games in a shop during the months of July and November.

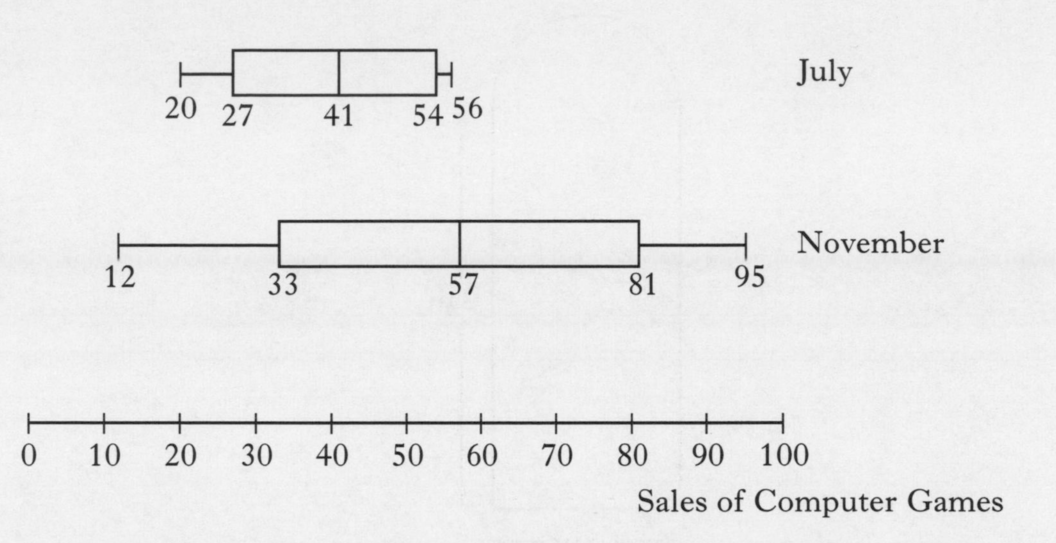

(*a*) Calculate the interquartile range for the November sales. **2**

(*b*) Explain how you can tell from the boxplots that the statement below is true.

"On average, the November sales are higher than July's but they also tend to be more variable." **2**

[Turn over for Question 12 on *Page fourteen*

DO NOT WRITE IN THIS MARGIN

Marks

12. The diagram below shows a window.

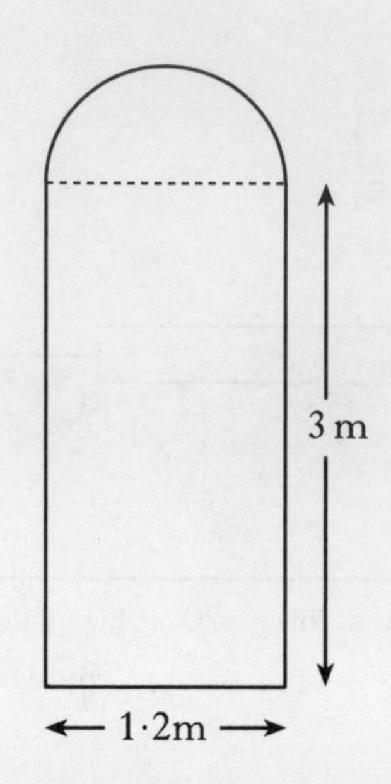

The window consists of a rectangle and a semi-circle.

Calculate the area of the window.

Give your answer in square metres correct to 2 decimal places.

5

[*END OF QUESTION PAPER*]

[BLANK PAGE]

FOR OFFICIAL USE

Total mark

X101/102

NATIONAL QUALIFICATIONS 2004

FRIDAY, 21 MAY
1.00 PM – 1.35 PM

MATHEMATICS
INTERMEDIATE 1
Units 1, 2 and
Applications of Mathematics
Paper 1
(Non-calculator)

Fill in these boxes and read what is printed below.

Full name of centre

Town

Forename(s)

Surname

Date of birth
Day Month Year

Scottish candidate number

Number of seat

1 **You may NOT use a calculator.**

2 Write your working and answers in the spaces provided. Additional space is provided at the end of this question-answer book for use if required. If you use this space, write clearly the number of the question involved.

3 Full credit will be given only where the solution contains appropriate working.

4 Before leaving the examination room you must give this book to the invigilator. If you do not you may lose all the marks for this paper.

LIB X101/102 6/6520

FORMULAE LIST

Circumference of a circle:	$C = \pi d$
Area of a circle:	$A = \pi r^2$
Curved surface area of a cylinder:	$A = 2\pi rh$

Theorem of Pythagoras:

c b a

$$a^2 + b^2 = c^2$$

DO NOT WRITE IN THIS MARGIN

Marks

ALL questions should be attempted.

1. Work out the answers to the following.

(*a*) 30% of £230 **1**

(*b*) $\frac{4}{7}$ of 105 **1**

(*c*) $380 - 20 \times 9$ **1**

2. A cooker can be bought by paying a deposit of £59 followed by 12 instalments of £45.

Calculate the total price of the cooker. **2**

[Turn over

DO NOT WRITE IN THIS MARGIN

Marks

3. Calculate the volume of this cuboid.

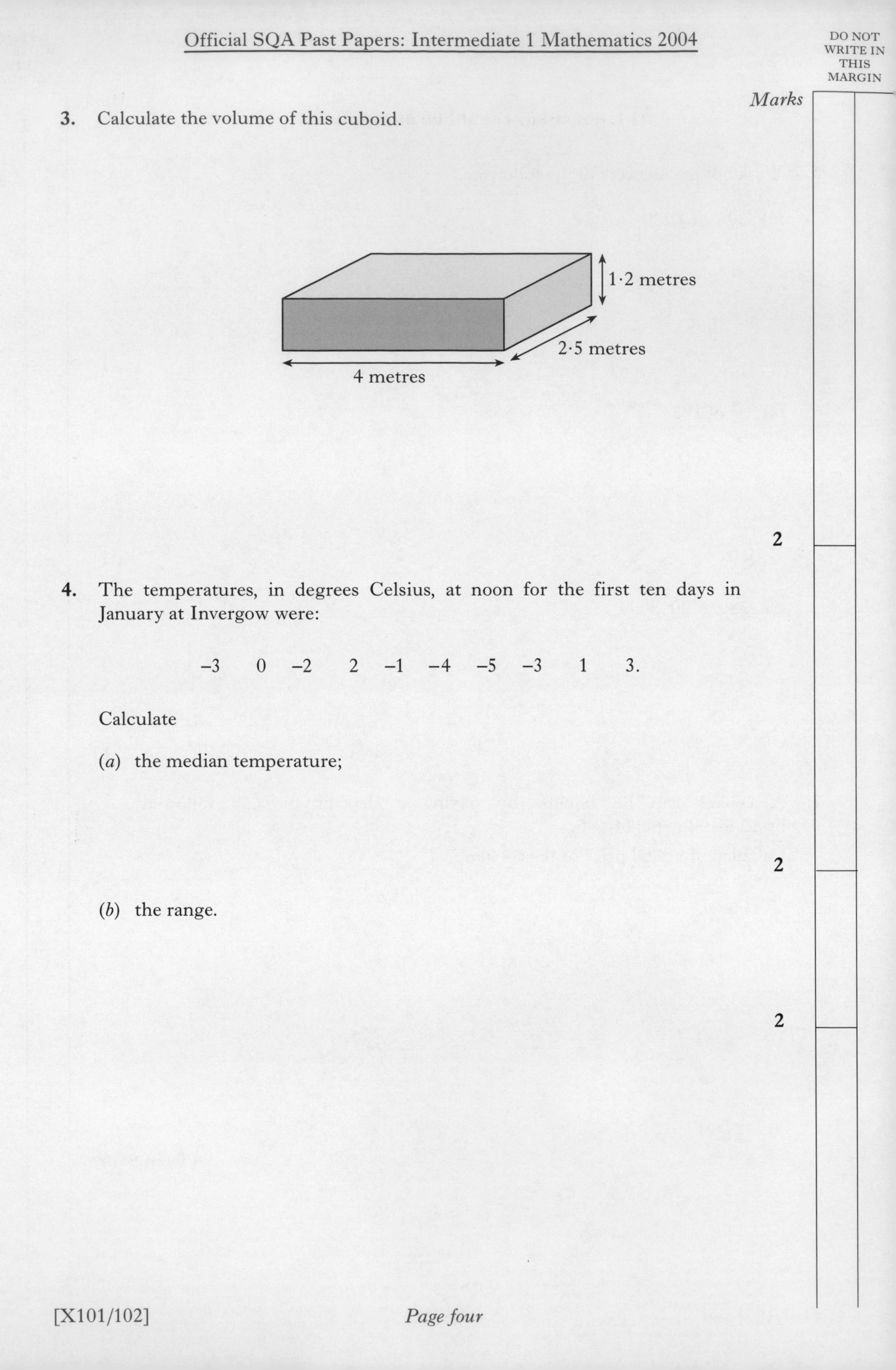

2

4. The temperatures, in degrees Celsius, at noon for the first ten days in January at Invergow were:

–3 0 –2 2 –1 –4 –5 –3 1 3.

Calculate

(*a*) the median temperature; **2**

(*b*) the range. **2**

DO NOT WRITE IN THIS MARGIN

Marks

4. (continued)

(*c*) The corresponding values of the median and the range for Abergrange are 2 °C and 5 °C respectively.

Make **two** comments comparing the temperatures in Invergow and Abergrange.

2

5. Farrah is a salesperson for an insurance company.

The flowchart below is used to work out her monthly salary in pounds.

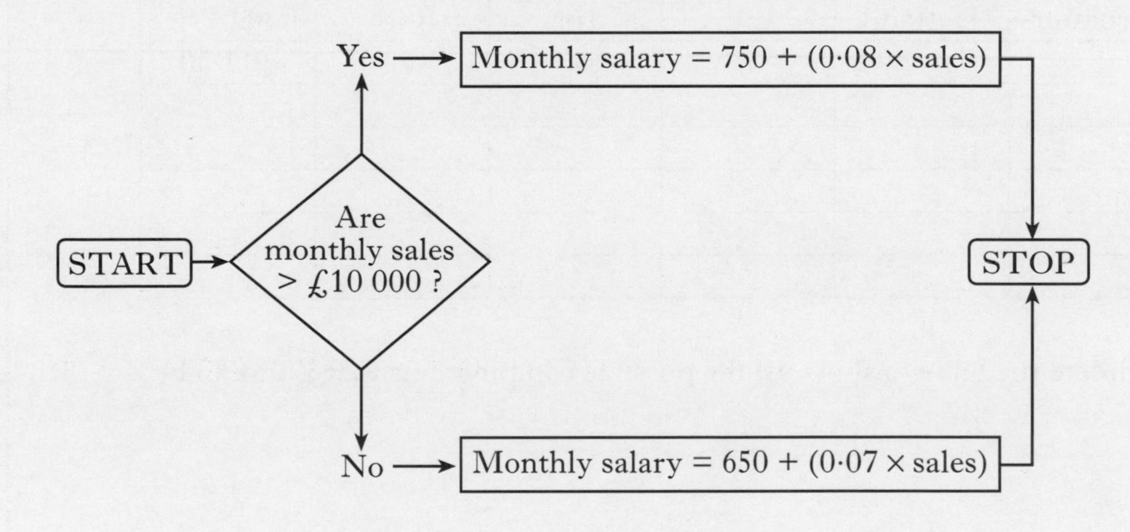

In March Farrah sold £9000 worth of insurance.

Calculate her salary for March.

2

[Turn over

DO NOT WRITE IN THIS MARGIN

Marks

6. A shop sells artificial flowers.
The prices of individual flowers are shown below.

Variety	**Price**
Carnation	£2
Daffodil	£3·50
Lily	£4
Iris	£3
Rose	£4·50

Zara wants to
- buy 3 flowers
- choose 3 different varieties
- spend a **minimum** of £10.

One combination of flowers that Zara can buy is shown in the table below.

Carnation	Daffodil	Lily	Iris	Rose	Total Price
		✓	✓	✓	£11·50

Complete the table to show **all** the possible combinations that Zara can buy. **3**

DO NOT WRITE IN THIS MARGIN

Marks

7. An Internet provider has a customer helpline.
The length of each telephone call to the helpline was recorded one day.
The results are shown in the frequency table below.

Length of call (to nearest minute)	Frequency	Length of call × Frequency
1	15	15
2	40	80
3	26	78
4	29	116
5	49	
6	41	
	Total = 200	Total =

(*a*) Complete the table above and find the mean length of call. **3**

(*b*) Write down the modal length of call. **1**

[Turn over

DO NOT WRITE IN THIS MARGIN

Marks

8. Brian is an apprentice plumber.

His basic rate of pay is £4·80 an hour for a 36 hour week.

His overtime rate of pay is time and a half.

Complete his payslip below for a week in which he works 5 hours overtime.

Payments				Deductions	
	Hours	Rate	Amount		Amount
Basic	36	£4·80	£172·80	Tax	£20·40
Overtime	5			National Insurance	£13·14
		Gross Pay		Total Deductions	£33·54
				Net Pay	

3

DO NOT WRITE IN THIS MARGIN

Marks

9. The scale drawing shows the positions of two towns, Allenby and Brucefield, which are 85 kilometres apart.

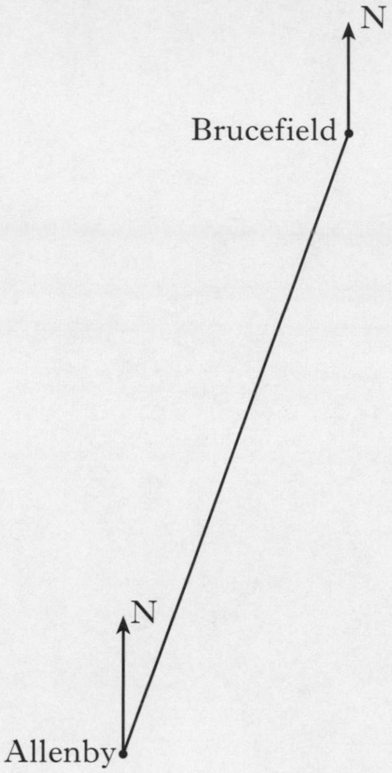

N

Brucefield

N

Allenby

(*a*) Find the scale of the drawing. **2**

(*b*) A third town, Cairnwell, is on a bearing of 060° from Allenby.
From Brucefield its bearing is 135°.
Complete the scale drawing above to show the position of Cairnwell. **3**

10. Evaluate $\frac{2xy}{z}$ when $x = -5$, $y = 6$ and $z = -4$. **3**

[*END OF QUESTION PAPER*]

DO NOT WRITE IN THIS MARGIN

ADDITIONAL SPACE FOR ANSWERS

FOR OFFICIAL USE

Total mark

X101/104

NATIONAL QUALIFICATIONS 2004

FRIDAY, 21 MAY
1.55 PM – 2.50 PM

MATHEMATICS
INTERMEDIATE 1
Units 1, 2 and
Applications of Mathematics
Paper 2

Fill in these boxes and read what is printed below.

Full name of centre

Town

Forename(s)

Surname

Date of birth
Day Month Year

Scottish candidate number

Number of seat

1 **You may use a calculator.**

2 Write your working and answers in the spaces provided. Additional space is provided at the end of this question-answer book for use if required. If you use this space, write clearly the number of the question involved.

3 Full credit will be given only where the solution contains appropriate working.

4 Before leaving the examination room you must give this book to the invigilator. If you do not you may lose all the marks for this paper.

FORMULAE LIST

Circumference of a circle: $C = \pi d$
Area of a circle: $A = \pi r^2$
Curved surface area of a cylinder: $A = 2\pi rh$

Theorem of Pythagoras:

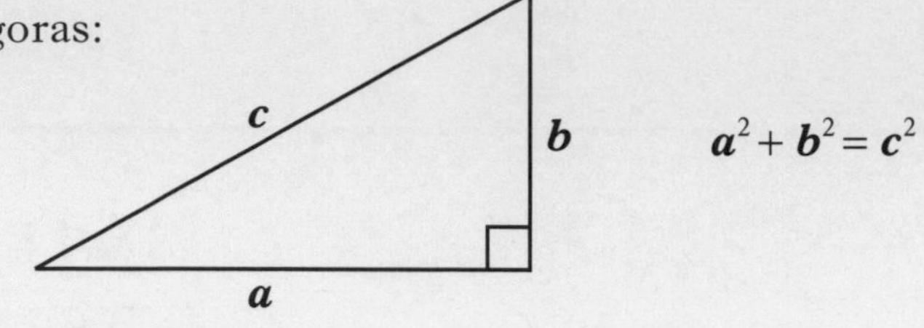

$a^2 + b^2 = c^2$

DO NOT WRITE IN THIS MARGIN

Marks

ALL questions should be attempted.

1. 2000 tickets are sold for a raffle in which the star prize is a television.
Kirsty buys 10 tickets for the raffle.
What is the probability that she wins the star prize?

1

2. (*a*) On the grid below, plot the points A(–3, 4), B(2, 4) and C(6, –5).

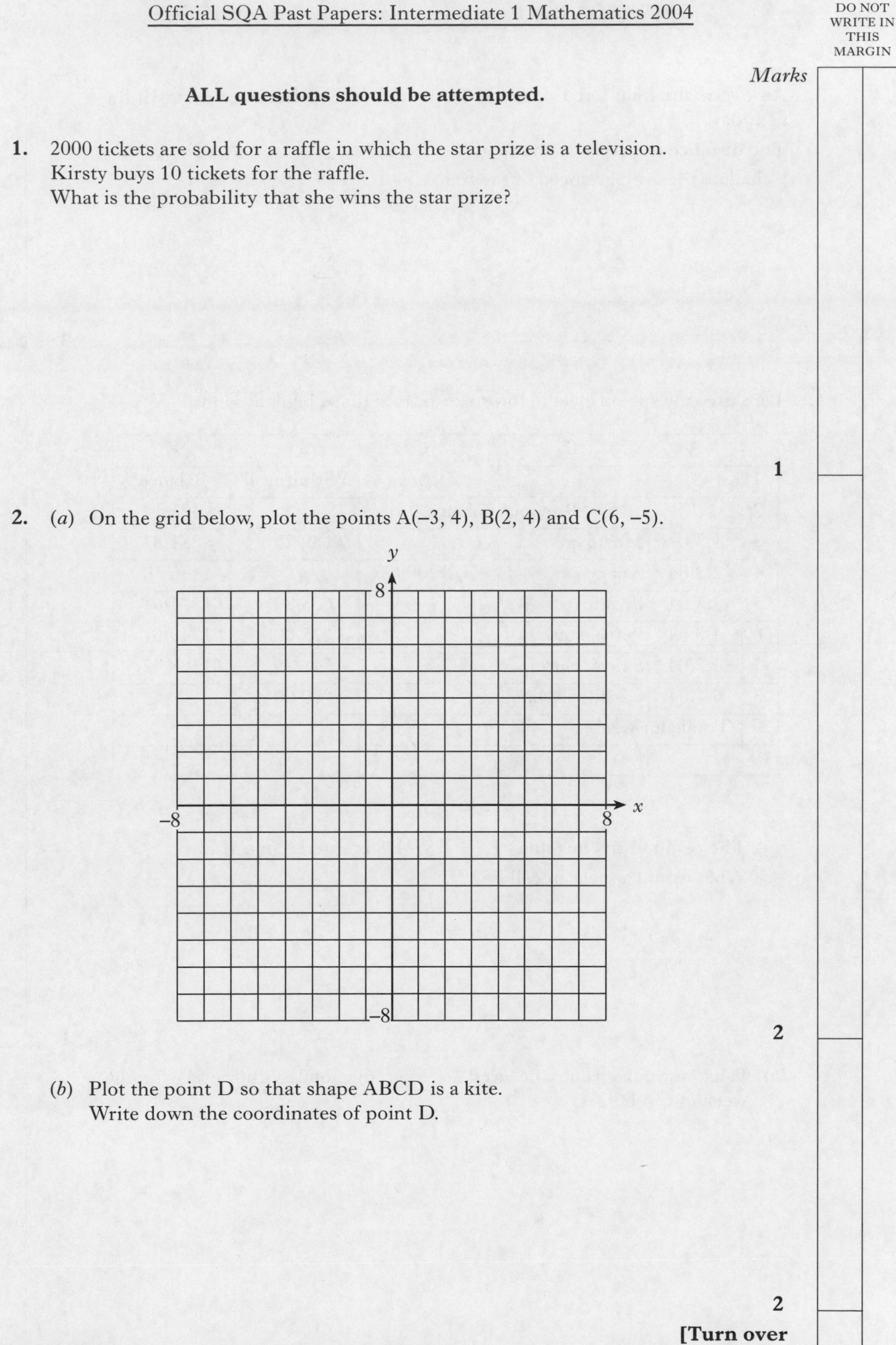

2

(*b*) Plot the point D so that shape ABCD is a kite.
Write down the coordinates of point D.

2

[Turn over

DO NOT WRITE IN THIS MARGIN

Marks

3. An overnight train left London at 2040 and reached Inverness at 0810 the next day.

The distance travelled by the train was 552 miles.

Calculate the average speed of the train. **3**

4. Tom uses the spreadsheet below to keep track of his bank account.

	A	B	C	D	E
1			Deposit	Withdrawal	Balance
2		Previous balance			£421·12
3	1.3.04	Mortgage		£189·25	£231·87
4	2.3.04	Salary	£1208·63		£1440·50
5	3.3.04	Presdos		£65·89	£1374·61
6	3.3.04	Petrol station		£25·00	£1349·61
7	4.3.04	Safe & Sure		£41·29	£1308·32
8	5.3.04	Cash from bank		£80·00	
9	Totals for week		£1208·63		
10					

(*a*) The result of the formula =E7 + C8 – D8 is entered in cell E8.

What would appear in cell E8? **1**

(*b*) What formula would be used to enter the total withdrawals for the week in cell D9? **1**

DO NOT WRITE IN THIS MARGIN

Marks

5. The scattergraph shows the age and mileage of cars in a garage.

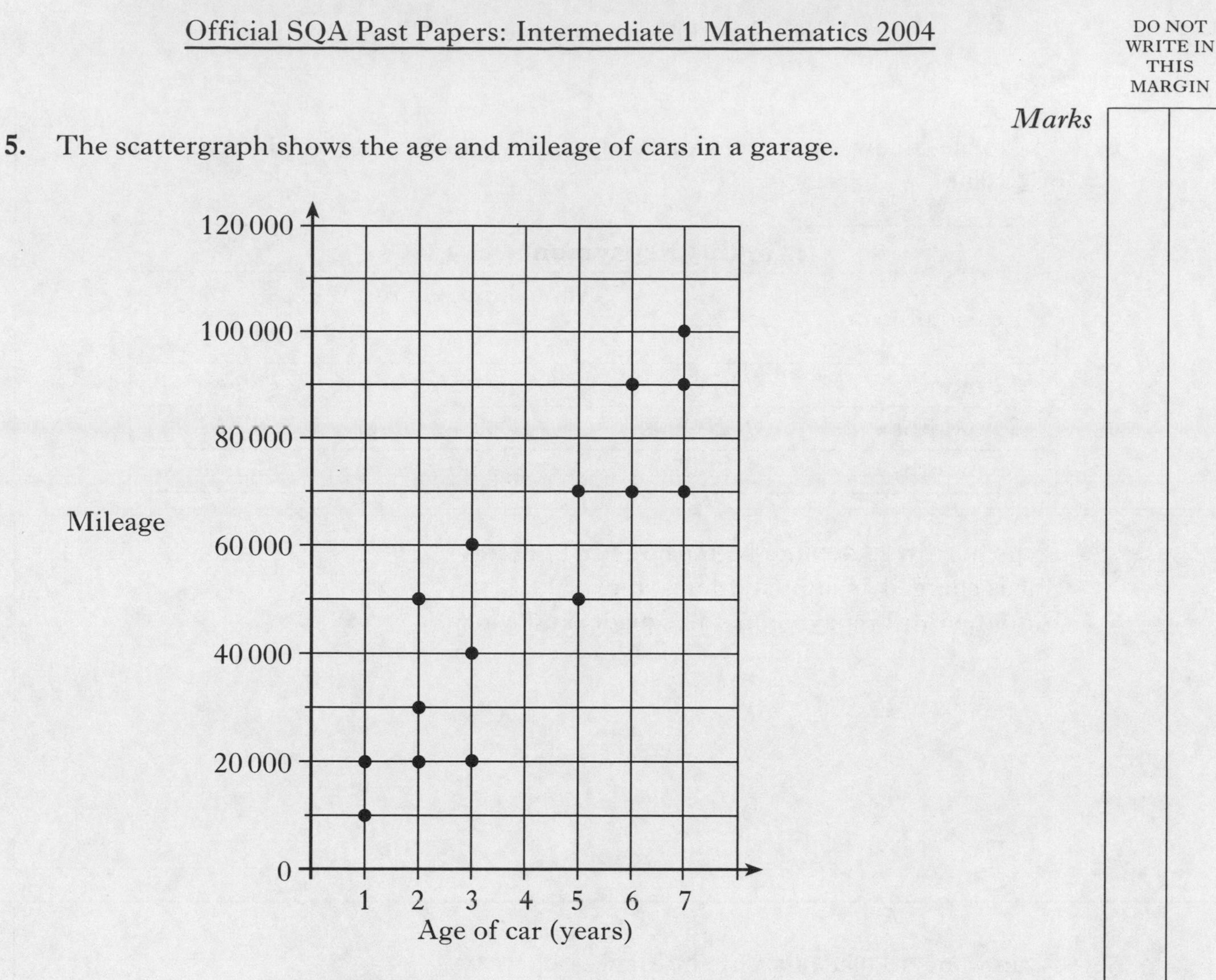

(*a*) Draw a line of best fit through the points on the graph. **1**

(*b*) Use your line of best fit to estimate the mileage of a 4 year old car. **1**

[Turn over

DO NOT WRITE IN THIS MARGIN

Marks

6. The table below shows the **monthly repayments** to be made on a loan of **£1000**.

Monthly Repayments on £1000			
Period of loan	Annual interest rate		
	10%	12%	14%
12 months	£87·72	£88·56	£89·40
24 months	£45·95	£46·79	£47·62
36 months	£32·07	£32·92	£33·78

Kylie borrows **£4000** to be repaid over 12 months.
She is charged an annual interest rate of 14%.
Find the **total** repayments to be made on the loan.

3

7. Ryan wants to take out a life insurance policy.
The insurance company charges a monthly premium of £2·50 for each £1000 of cover.
Ryan can afford to pay £90 per month.
How much cover can he get?

2

DO NOT WRITE IN THIS MARGIN

Marks

8. (*a*) In a jewellery shop the price of a gold chain is proportional to its length.
A 16 inch gold chain is priced at £40.
Calculate the price of a 24 inch gold chain. **2**

(*b*) The gold chains are displayed diagonally on a **square** board of side 20 inches.
The longest chain stretches from corner to corner.
Calculate the length of the longest chain.
Do not use a scale drawing.

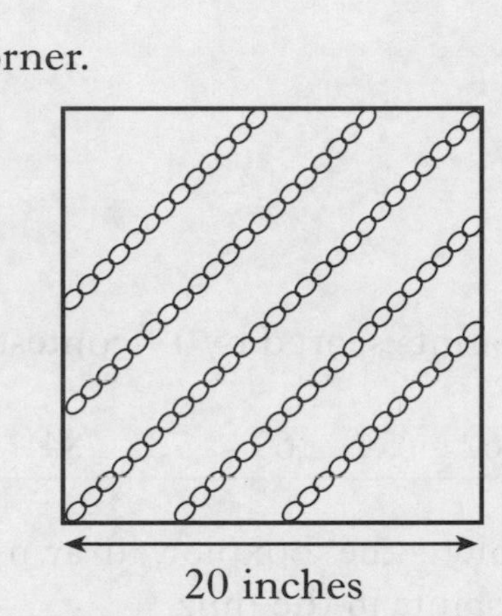

3

[Turn over

DO NOT WRITE IN THIS MARGIN

Marks

9. Andy buys a bottle of aftershave in Spain for 38·50 euros.
The same bottle of aftershave costs £25·99 in Scotland.
The exchange rate is £1 = 1·52 euros.
Does he save money by buying the aftershave in Spain?
Explain your answer.

3

10. The points scored by 14 contestants in a quiz are shown below.

60 62 54 69 73 84 83 75 91 77 93 88 70 72

Complete the boxplot, drawn below, to show the points scored by the contestants in the quiz.

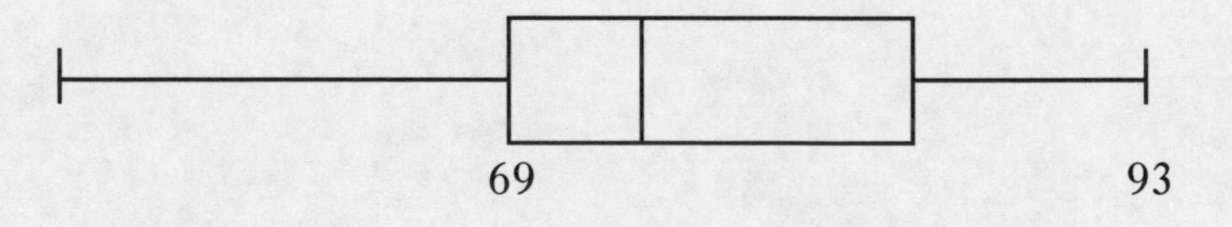

4

DO NOT WRITE IN THIS MARGIN

Marks

11. The graph below shows how the rate of interest for a savings account with the Clydeside Bank changed during 2002.

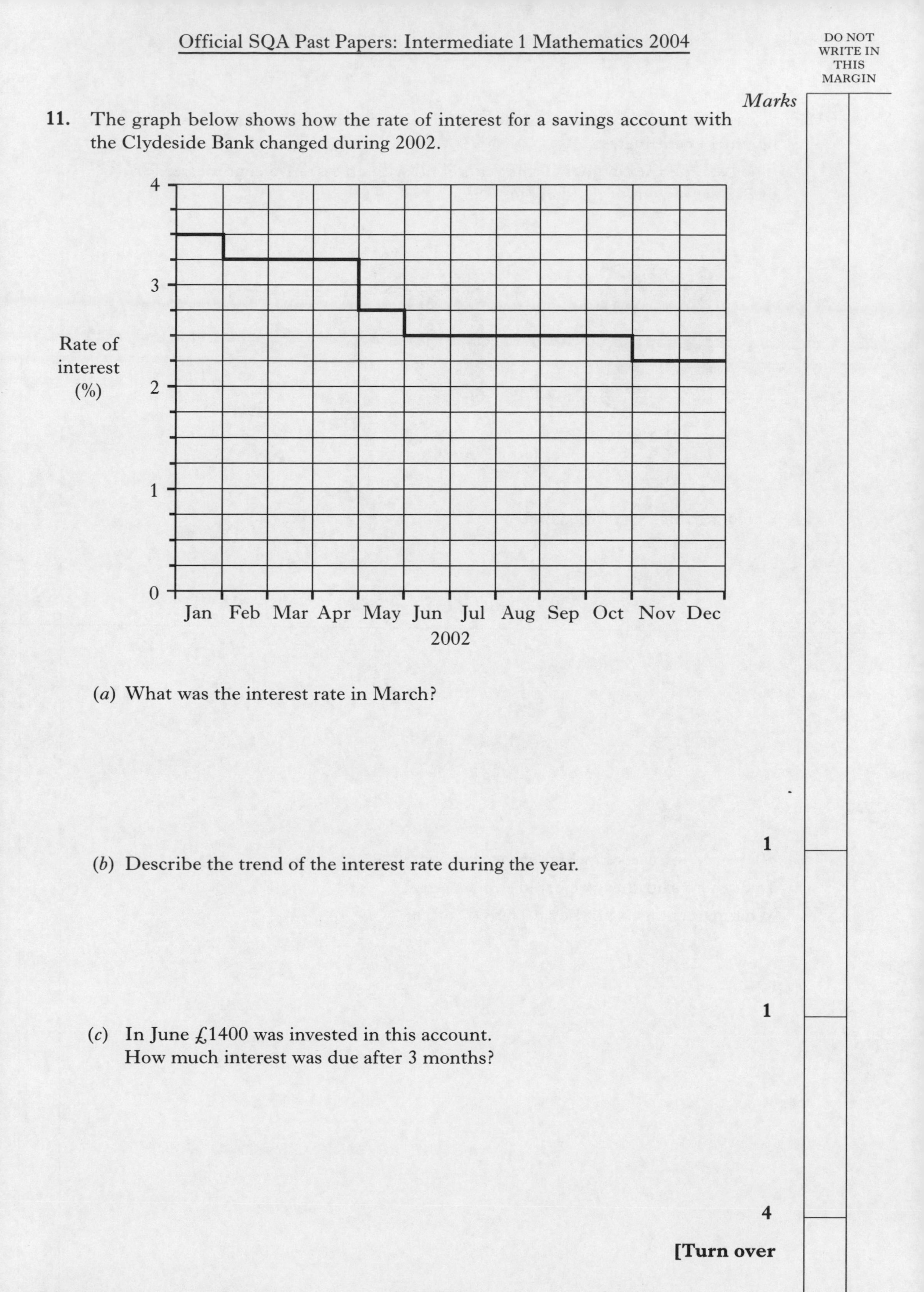

(*a*) What was the interest rate in March? **1**

(*b*) Describe the trend of the interest rate during the year. **1**

(*c*) In June £1400 was invested in this account.
How much interest was due after 3 months? **4**

[Turn over

DO NOT WRITE IN THIS MARGIN

Marks

12. A soup tin is in the shape of a cylinder with diameter 10 centimetres and height 14 centimetres.

The label on the tin goes all the way round with an extra 1·5 centimetres for overlap as shown in the diagram.

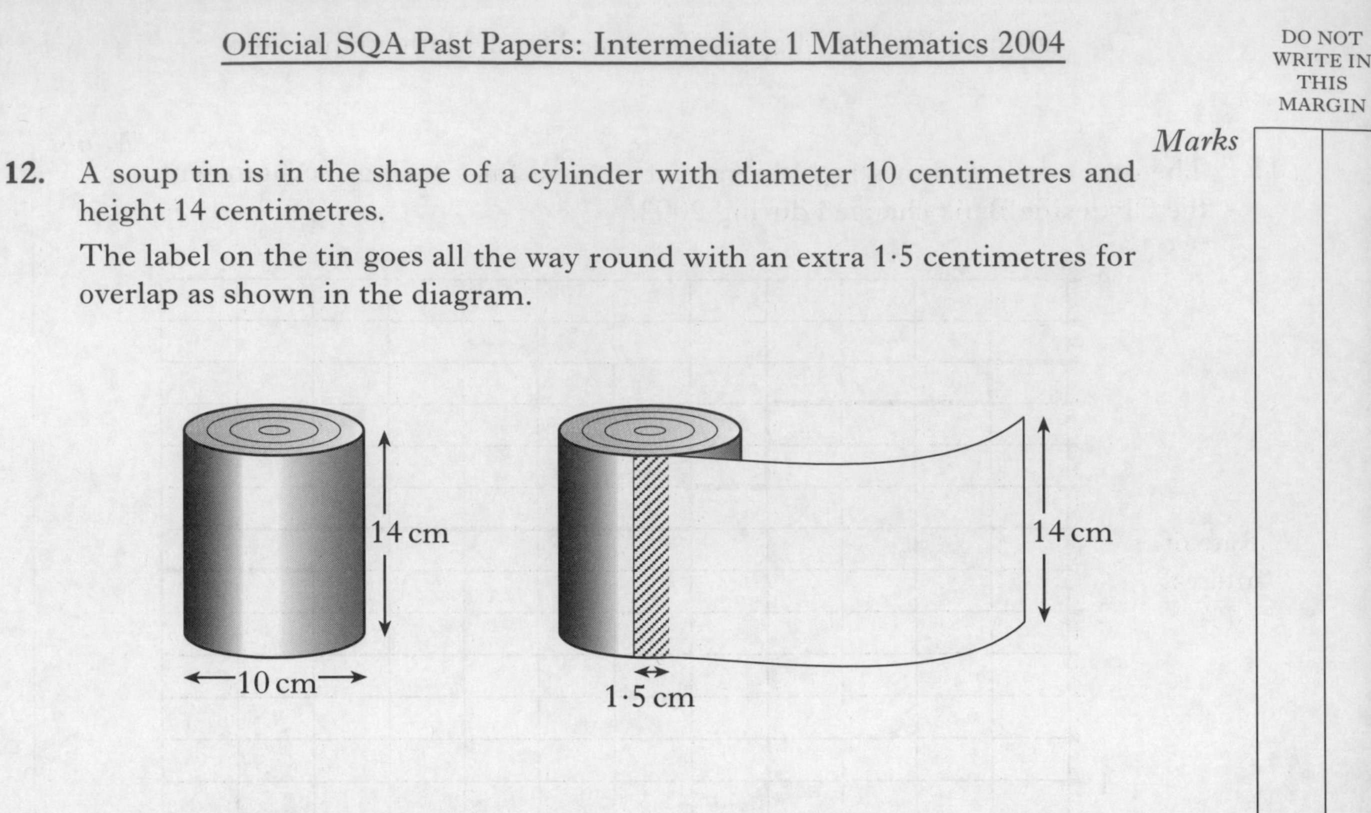

Calculate the area of the label. **4**

13. 40 people were asked whether they preferred tea or coffee.

18 of them said they preferred coffee.

What percentage said they preferred coffee? **3**

DO NOT WRITE IN THIS MARGIN

Marks

14. The diagram below shows a rectangular door with a window.

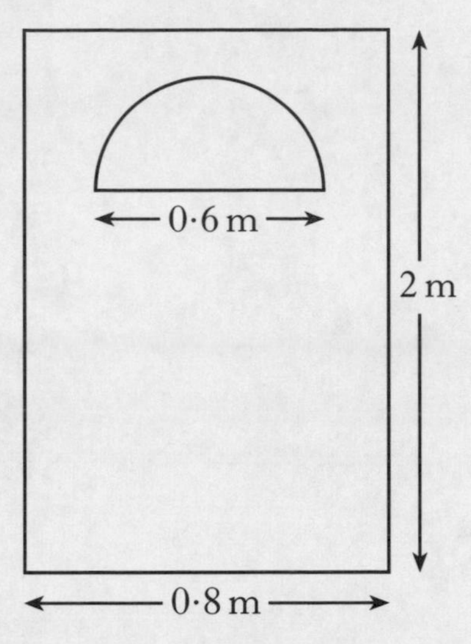

The window is in the shape of a semi-circle and is made of glass.

The rest of the door is made of wood.

Calculate the area of the wooden part of the door.

Give your answer in square metres correct to two decimal places.

5

[*END OF QUESTION PAPER*]

DO NOT WRITE IN THIS MARGIN

ADDITIONAL SPACE FOR ANSWERS

[BLANK PAGE]

FOR OFFICIAL USE

Total mark

X101/102

NATIONAL QUALIFICATIONS 2005

FRIDAY, 20 MAY
1.00 PM – 1.35 PM

MATHEMATICS
INTERMEDIATE 1
Units 1, 2 and Applications of Mathematics
Paper 1
(Non-calculator)

Fill in these boxes and read what is printed below.

Full name of centre

Town

Forename(s)

Surname

Date of birth
Day Month Year

Scottish candidate number

Number of seat

1 **You may <u>NOT</u> use a calculator.**

2 Write your working and answers in the spaces provided. Additional space is provided at the end of this question-answer book for use if required. If you use this space, write clearly the number of the question involved.

3 Full credit will be given only where the solution contains appropriate working.

4 Before leaving the examination room you must give this book to the invigilator. If you do not you may lose all the marks for this paper.

FORMULAE LIST

Circumference of a circle: $C = \pi d$

Area of a circle: $A = \pi r^2$

Curved surface area of a cylinder: $A = 2\pi rh$

Theorem of Pythagoras:

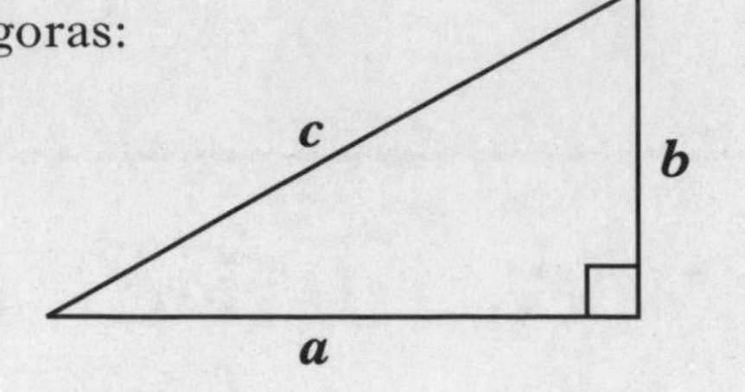

$$a^2 + b^2 = c^2$$

DO NOT WRITE IN THIS MARGIN

Marks

ALL questions should be attempted.

1. (*a*) Find 6·17 – 2·3. **1**

(*b*) Find 75% of £1200. **1**

(*c*) Write $\frac{7}{1000}$ as a decimal. **1**

2. Joyce is going on holiday. She must be at the airport by 1.20 pm. It takes her 4 hours 30 minutes to travel from home to the airport. What is the latest time that she should leave home for the airport? **1**

[Turn over

DO NOT WRITE IN THIS MARGIN

Marks

3. A regular polygon is a shape with three or more equal sides.

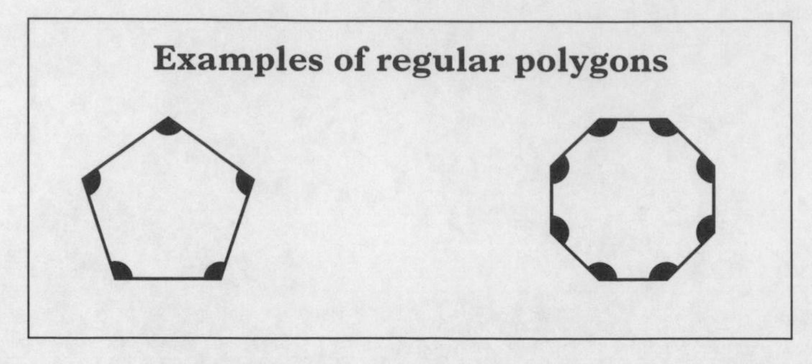

A rule used to calculate the size, in degrees, of each angle in a regular polygon is:

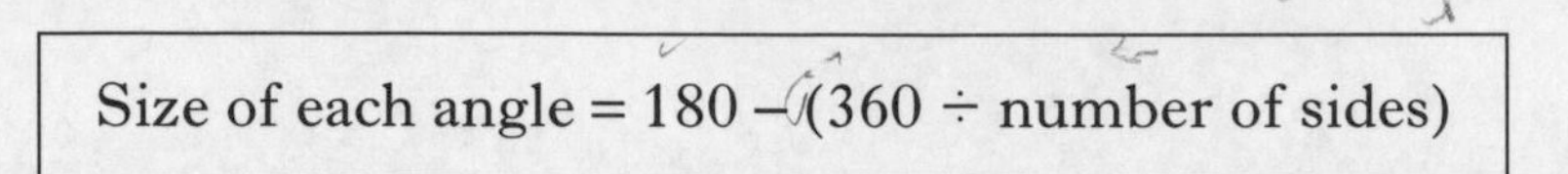

Calculate the size of each angle in the regular polygon below.

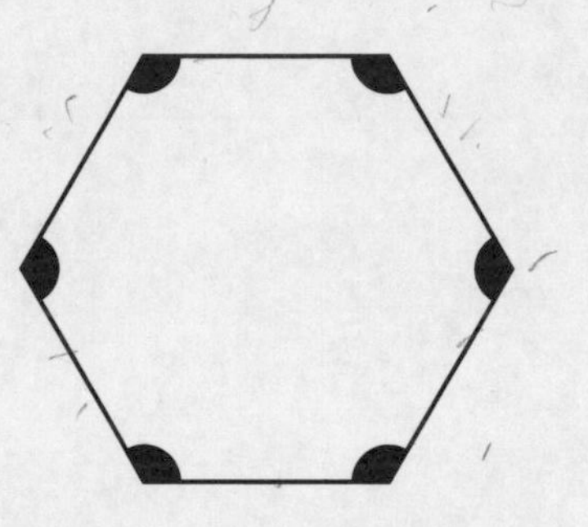

Do not measure with a protractor.

You must show your working.

2

DO NOT WRITE IN THIS MARGIN

Marks

4. The number of peas counted in each of 100 pea pods is shown in this frequency table.

Peas in pod	Frequency	Peas in pod × Frequency
3	5	15
4	10	40
5	28	140
6	36	216
7	12	
8	9	
	Total = 100	Total =

Complete the table above **and** calculate the mean number of peas in a pod.

3

[Turn over

DO NOT WRITE IN THIS MARGIN

Marks

5. The network diagram shows the streets in Jo's newspaper delivery round.

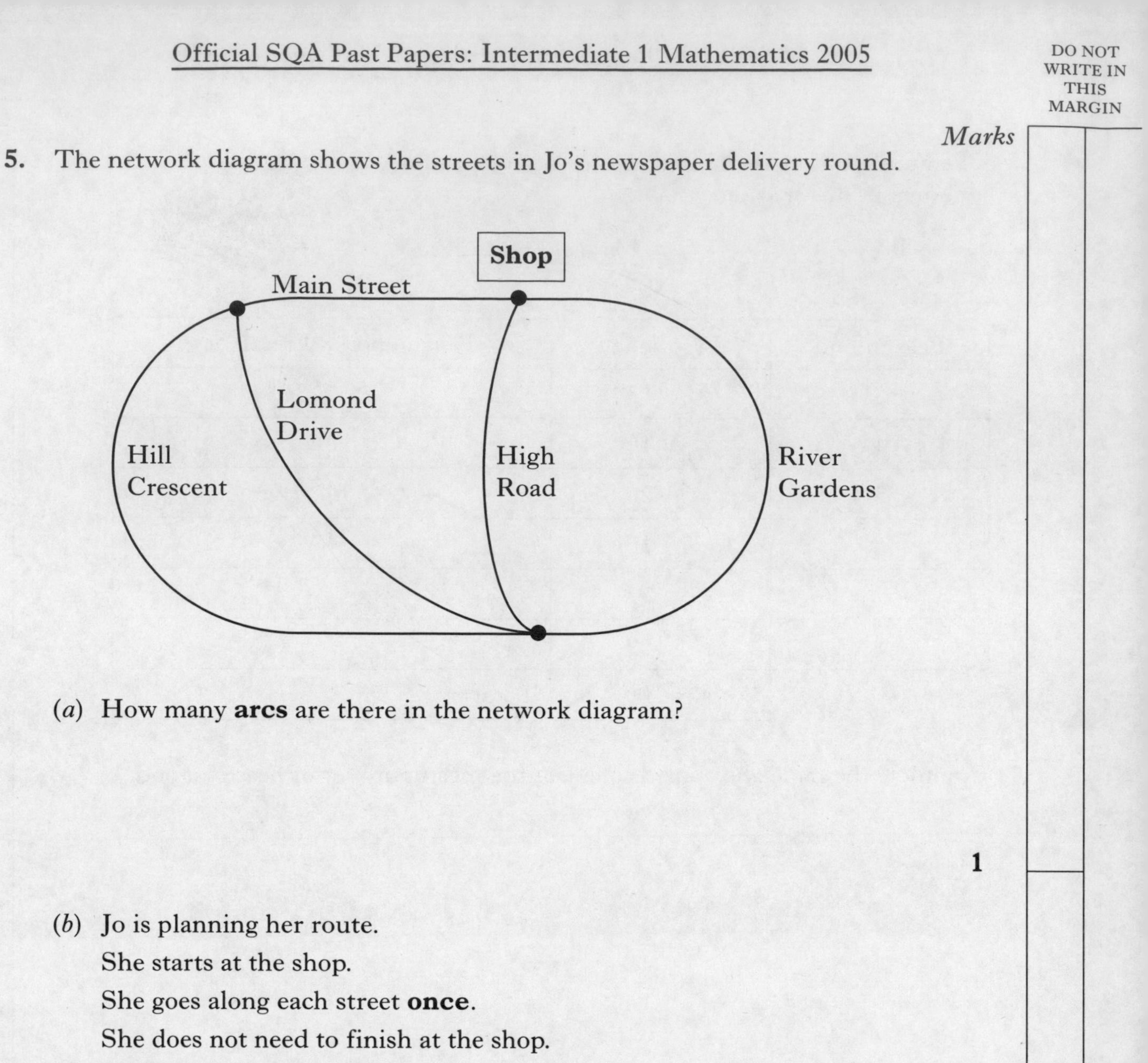

(*a*) How many **arcs** are there in the network diagram?

1

(*b*) Jo is planning her route.
She starts at the shop.
She goes along each street **once**.
She does not need to finish at the shop.
List the streets, in order, for **one** possible route.

1

Marks

6. Anwar wants to buy some accessories for his computer.
He sees this advert for Cathy's Computers.

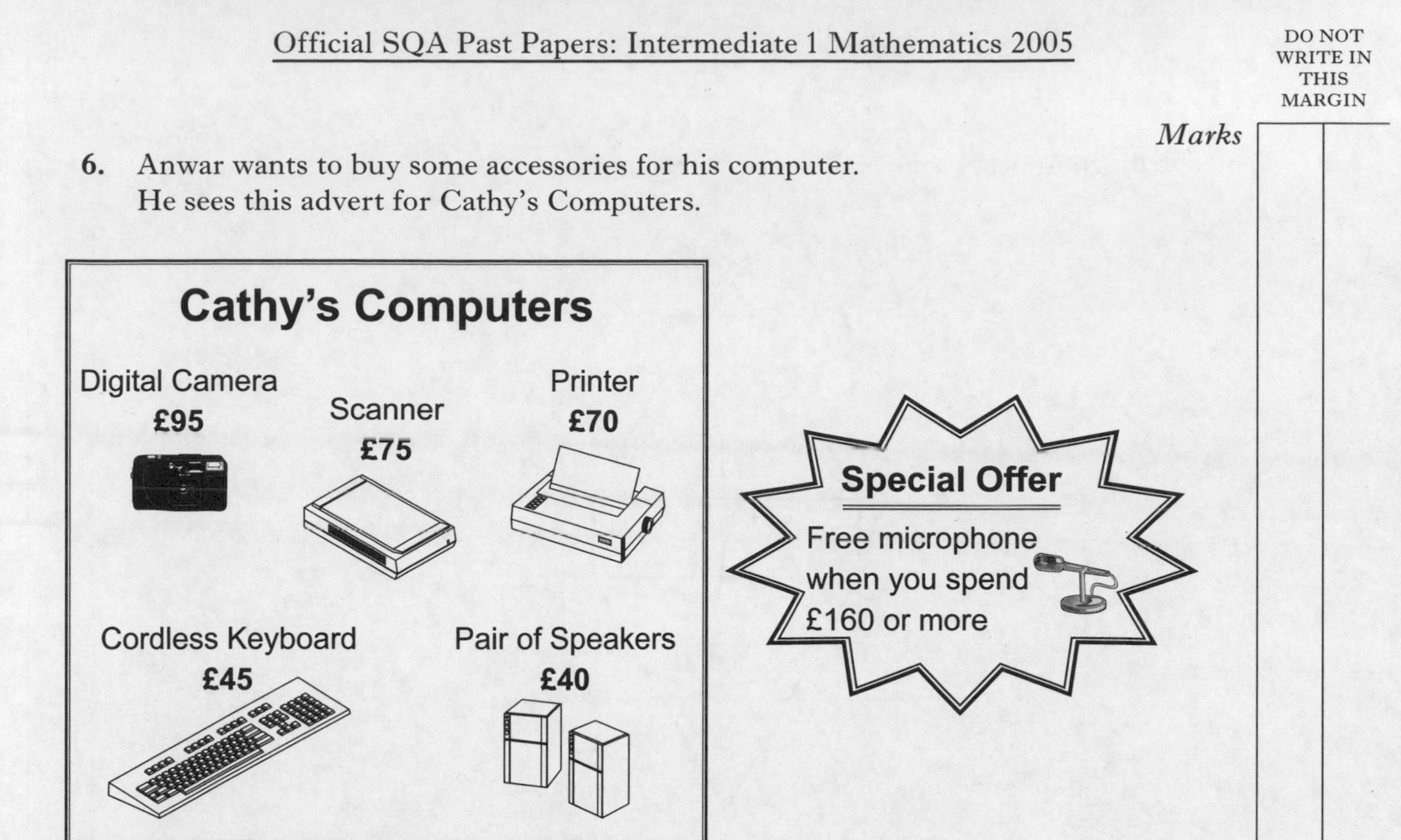

Anwar wants to spend enough to get the free microphone.
He can afford to spend a maximum of £200.
He does not want to buy more than one of each accessory.

One combination of accessories that Anwar can buy is shown in the table below.

Digital Camera £95	Scanner £75	Printer £70	Cordless Keyboard £45	Pair of Speakers £40	Total Value
	✓	✓		✓	£185

Complete the table to show **all** possible combinations that Anwar can buy. **3**

[Turn over

DO NOT WRITE IN THIS MARGIN

Marks

7. The diagram below shows the net of a triangular prism.

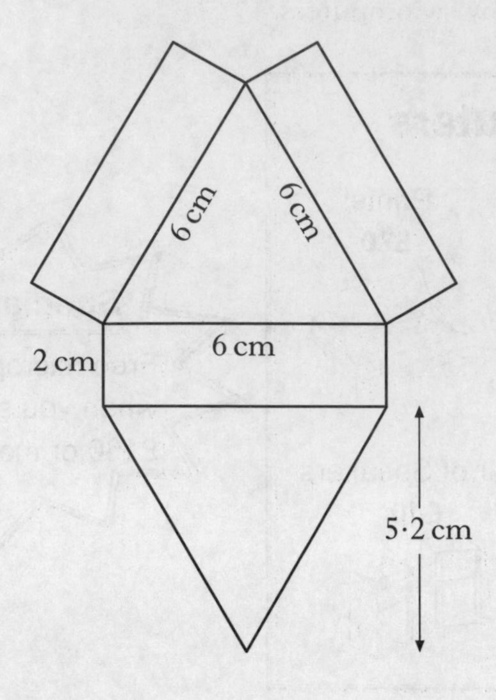

Find the total **surface area** of the triangular prism.

3

DO NOT WRITE IN THIS MARGIN

Marks

8. (*a*) While in New York, Martin changed £50 into US dollars.

The exchange rate was £1 = $1·62.

How many US dollars did Martin receive for £50? **2**

(*b*) A few days later he received $320 in exchange for £200.

What was the new exchange rate? **2**

[Turn over

DO NOT WRITE IN THIS MARGIN

Marks

9. The scale drawing shows some streets in a town.

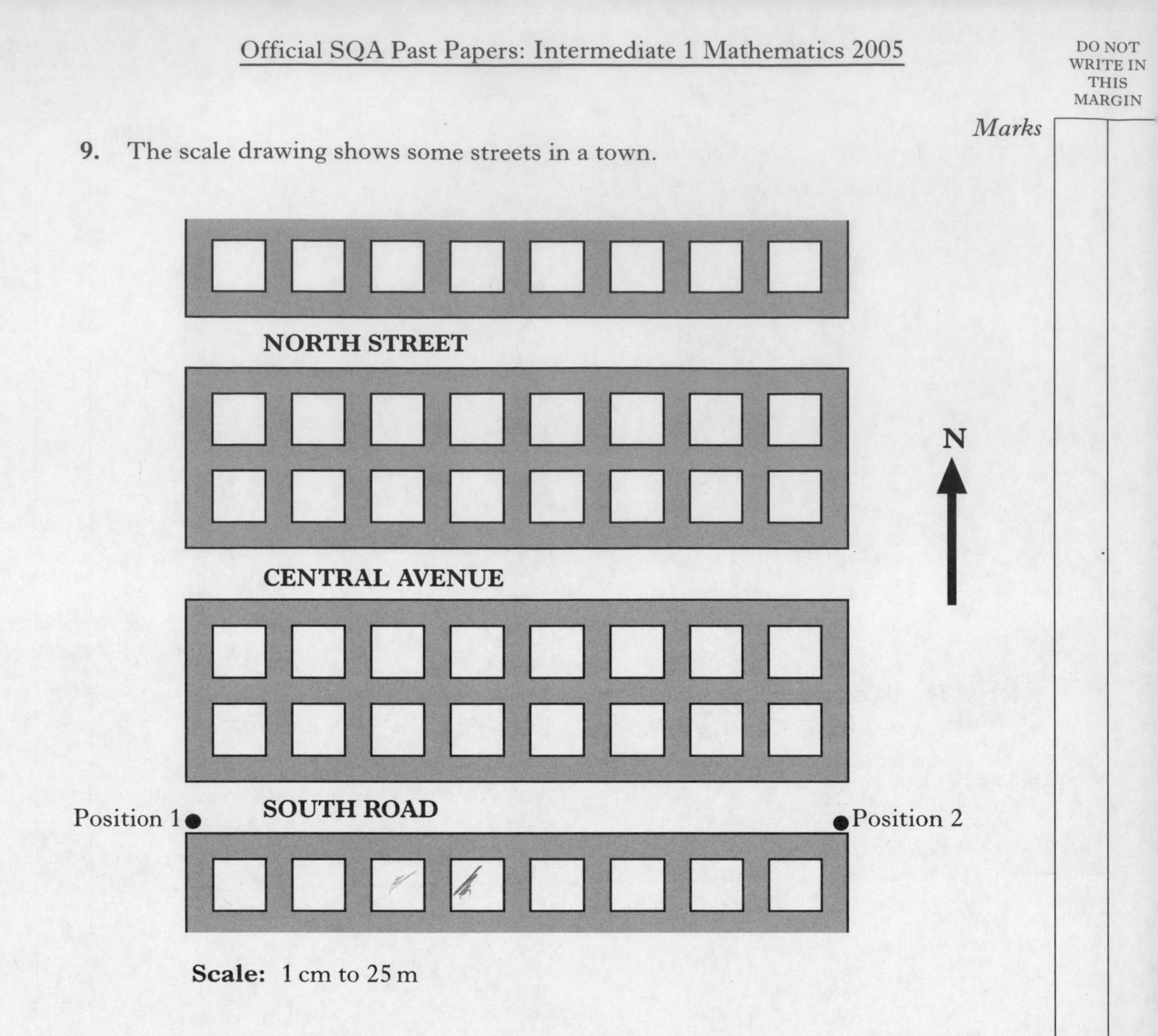

(*a*) Use the scale drawing to find the length, in metres, of North Street. **2**

(*b*) A television detector van drives along South Road.
It detects a television being used without a licence on a bearing of

- 050° from Position 1
- 320° from Position 2.

Complete the scale drawing above to show the position of the house where the television is being used without a licence. **3**

DO NOT WRITE IN THIS MARGIN

Marks

10. In a **magic square**, the numbers in each row, each column and each diagonal add up to the same **magic total**.

In this magic square the **magic total** is 3.

–2	5	0
3	1	–1
2	–3	4

(*a*)

–4	3	–2
1	–1	–3
0	–5	2

This is another magic square.
What is its **magic total**? **1**

(*b*) Complete this **magic square**. **3**

1		
	–2	
–3		–5

[*END OF QUESTION PAPER*]

DO NOT WRITE IN THIS MARGIN

ADDITIONAL SPACE FOR ANSWERS

FOR OFFICIAL USE

Total mark

X101/104

NATIONAL QUALIFICATIONS 2005

FRIDAY, 20 MAY
1.55 PM – 2.50 PM

MATHEMATICS
INTERMEDIATE 1
Units 1, 2 and
Applications of Mathematics
Paper 2

Fill in these boxes and read what is printed below.

Full name of centre

Town

Forename(s)

Surname

Date of birth
Day Month Year

Scottish candidate number

Number of seat

1 **You may use a calculator.**

2 Write your working and answers in the spaces provided. Additional space is provided at the end of this question-answer book for use if required. If you use this space, write clearly the number of the question involved.

3 Full credit will be given only where the solution contains appropriate working.

4 Before leaving the examination room you must give this book to the invigilator. If you do not you may lose all the marks for this paper.

LIB X101/104 6/4170

FORMULAE LIST

Circumference of a circle:	$C = \pi d$
Area of a circle:	$A = \pi r^2$
Curved surface area of a cylinder:	$A = 2\pi rh$

Theorem of Pythagoras:

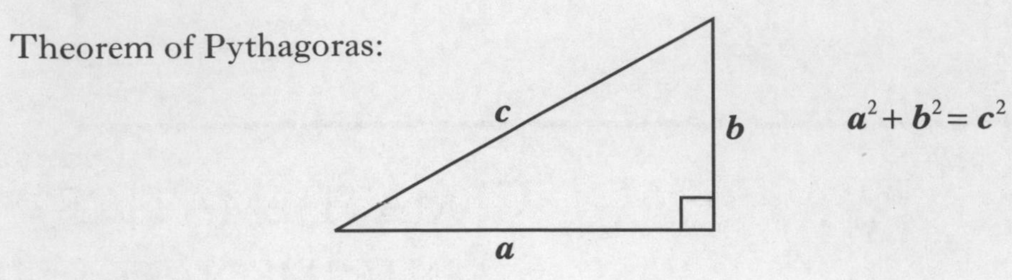

$$a^2 + b^2 = c^2$$

DO NOT WRITE IN THIS MARGIN

Marks

ALL questions should be attempted.

1. Calculate the volume of the cube below.

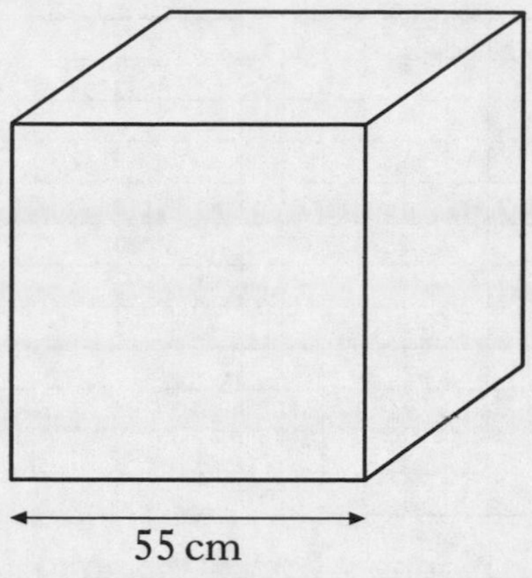

55 cm

Round your answer to the nearest thousand cubic centimetres. **2**

2. Claire sells cars.
She is paid £250 per month plus 3% commission on her sales.
How much is she paid in a month when her sales are worth £72 000? **2**

[Turn over

DO NOT WRITE IN THIS MARGIN

Marks

3. A group of students visit a theme park.

The graph below shows their journey.

They set off from the college at 9 am and arrive back at 4 pm.

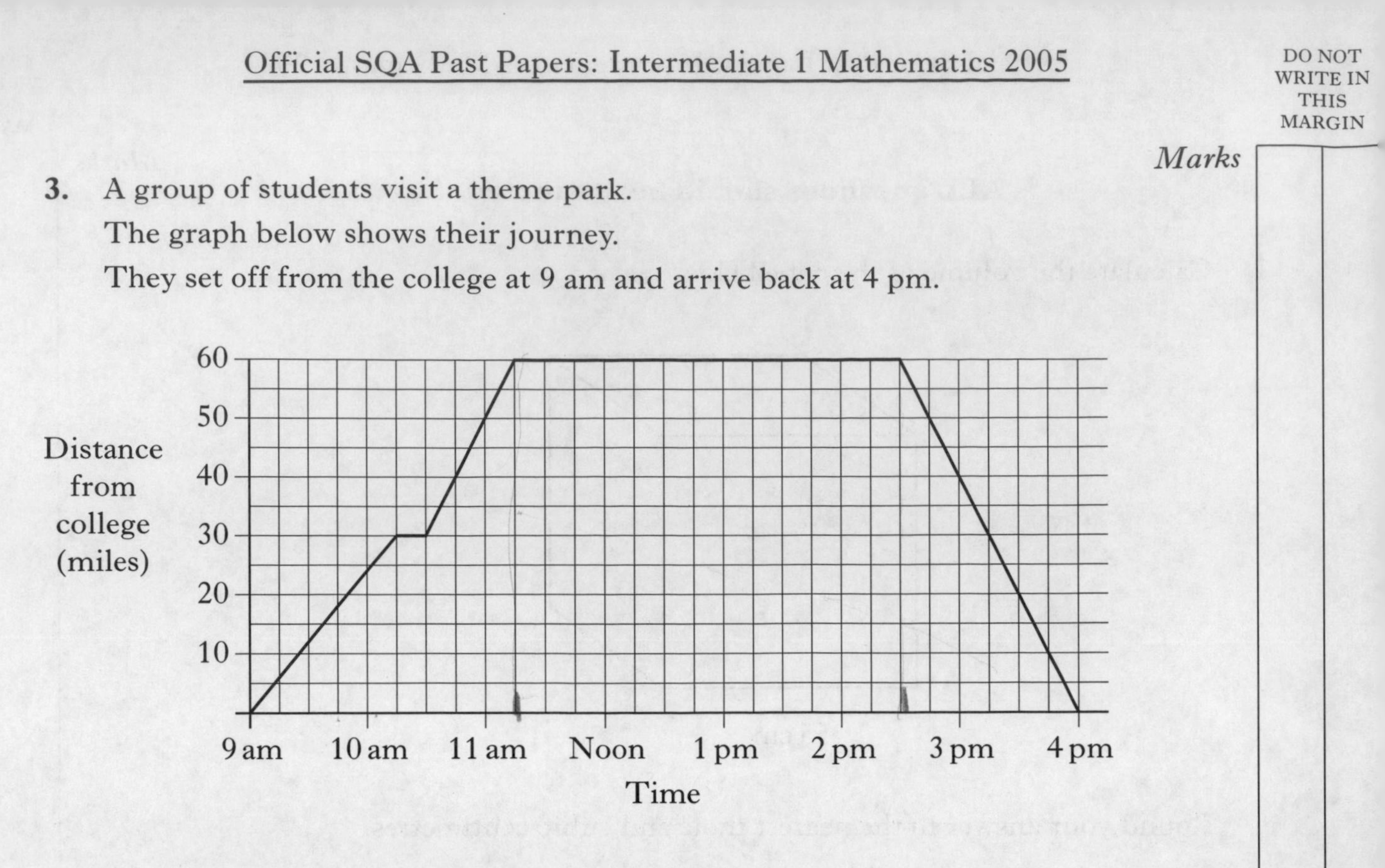

(*a*) How long did the students spend at the theme park? **1**

(*b*) Calculate the average speed, in miles per hour, of the students' return journey. **3**

DO NOT WRITE IN THIS MARGIN

Marks

4. A superstore uses a spreadsheet to work out its weekly wage bill.

	A	B	C	D
1	Employee	Weekly	Number of	Total Weekly
2	Grade	Wage in £	Employees	Wage in £
3				
4	1	140	10	1400
5	2	155	25	3875
6	3	175	31	5425
7	4	205	38	7790
8	5	240	17	4080
9	6	280	14	3920
10	7	330	9	
11				
12		Total Weekly Wage Bill		
13				

(*a*) What formula would be used to enter the total weekly wage for grade 7 employees in cell D10? **1**

(*b*) What formula would be used to enter the total weekly wage bill in cell D12? **1**

[Turn over

DO NOT WRITE IN THIS MARGIN

Marks

5. The stem and leaf diagram below shows the ages of the players in the Kestrels rugby team.

AGES
Kestrels

Stem	Leaf
1	9
2	1 3 4 7 9
3	0 2 4 5 5 5 8 9
4	1

2 | 1 represents 21 years

(*a*) What age is the oldest player? **1**

(*b*) Calculate the range of ages. **2**

The stem and leaf diagram below shows the ages of both the Kestrels and the Falcons rugby teams.

AGES

Falcons	Stem	Kestrels
9 9	1	9
8 7 7 6 3 2 1 1 0	2	1 3 4 7 9
8 6 4 3	3	0 2 4 5 5 5 8 9
	4	1

2 | 1 represents 21 years

(*c*) Compare the ages of the two teams. Comment on any difference. **1**

DO NOT WRITE IN THIS MARGIN

Marks

6. The table below shows the **monthly payments** to be made when money is borrowed from a finance company.

Money can be borrowed with (**w**) or without (**w/o**) loan protection.

		Amount Borrowed		
		£5000	**£8000**	**£10 000**
Term		Monthly Payment		
24 months	**w**	£246·67	£392·38	£490·84
	w/o	£222·81	£357·40	£447·12
36 months	**w**	£170·58	£273·83	£342·67
	w/o	£153·34	£246·24	£308·18
48 months	**w**	£136·63	£219·52	£274·75
	w/o	£118·69	£190·81	£238·88
60 months	**w**	£114·49	£184·09	£230·49
	w/o	£97·97	£157·66	£197·45

(*a*) Harry borrows £8000 over 4 years **with loan protection**. State his monthly payment. **1**

(*b*) Over the 4 years, how much **less** would Harry pay **in total** if he borrowed £8000 without loan protection? **2**

[Turn over

DO NOT WRITE IN THIS MARGIN

Marks

7. The scores of 12 golfers in a competition were as follows.

67	70	68	75	71	70
70	75	76	75	74	75

(*a*) Find the modal score. **1**

(*b*) Find the median score. **2**

(*c*) Find the probability of choosing a golfer from this group with a score of 70. **1**

8. 60 workers in a factory voted on a new pay deal.
42 of them voted to accept the deal.
What percentage voted to accept the deal? **3**

DO NOT WRITE IN THIS MARGIN

Marks

9. The pie chart shows the different sizes of eggs laid by a flock of hens.

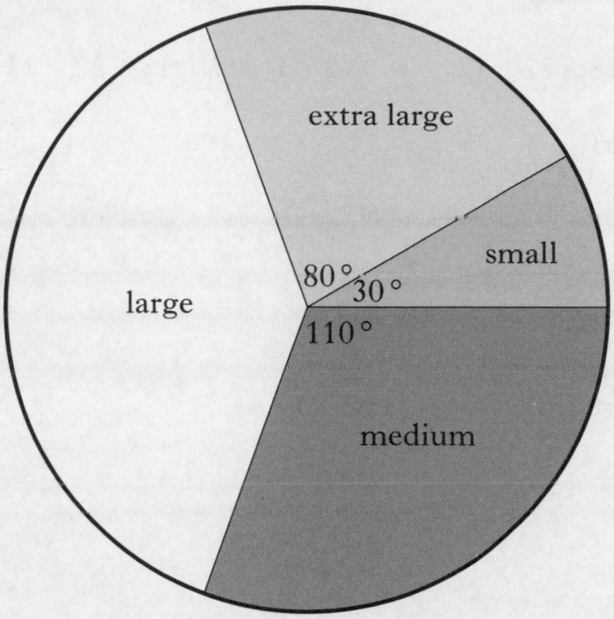

The flock of hens laid 1260 eggs.
How many of the eggs were large?

3

[Turn over

DO NOT WRITE IN THIS MARGIN

Marks

10. A group of 17 people were asked to state the number of hours they spent watching television one week.

Their replies are shown below.

2 5 8 9 11 7 28 27 12 10 12 1 13 15 22 23 2

(*a*) Find the lower quartile. **2**

(*b*) Calculate the interquartile range. **2**

11. The diagram below shows a speedway track.

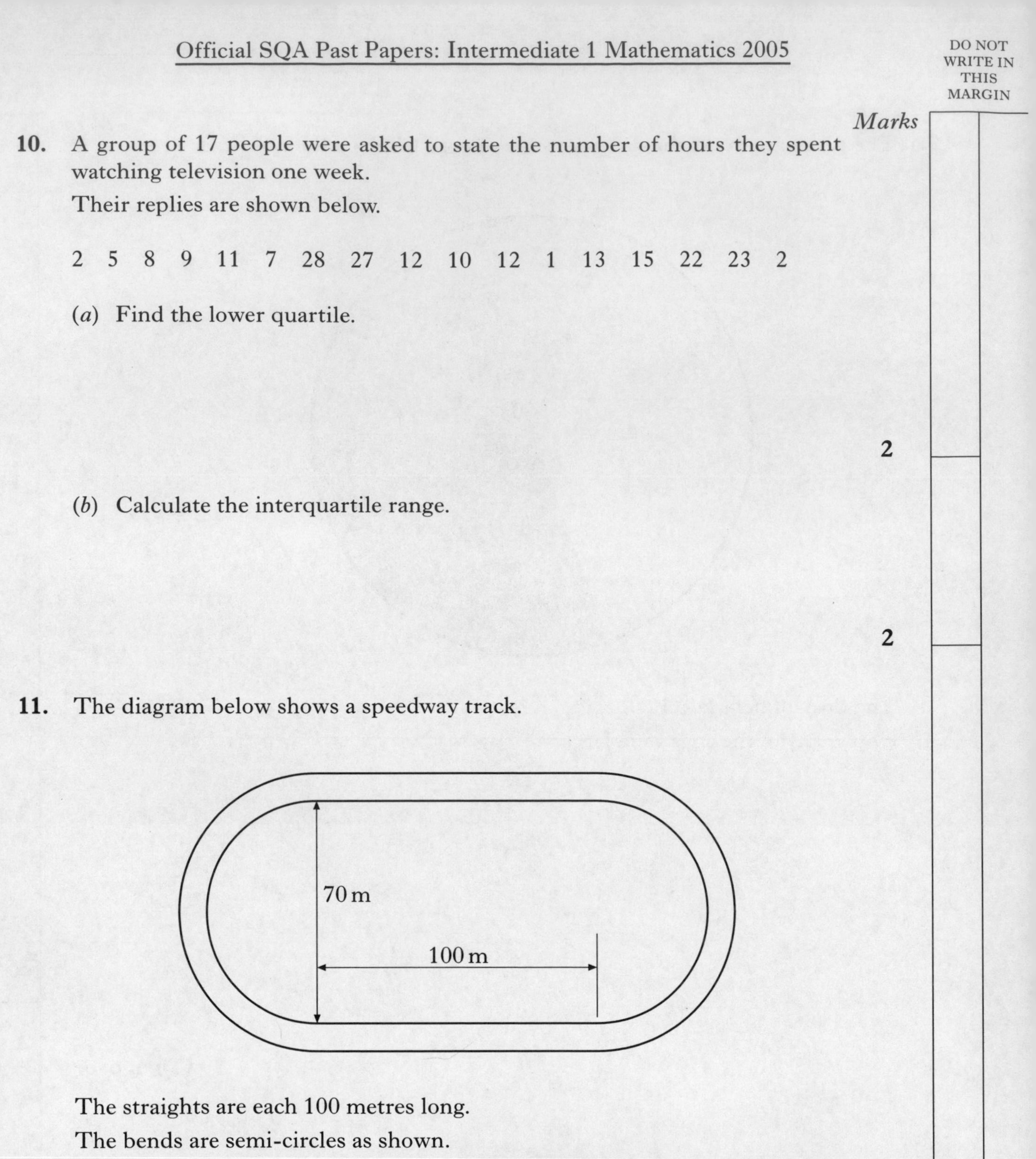

The straights are each 100 metres long.

The bends are semi-circles as shown.

Calculate the perimeter of the inside of the track. **4**

DO NOT WRITE IN THIS MARGIN

Marks

12. The advert below for a receptionist was displayed in a job centre.

RECEPTIONIST

Hours:	Mon–Thu	0800–1200 and 1300–1630
	Fri	0800–1200
	Sat	0800–1200
Pay:	Mon–Fri	£6·50 per hour
	Sat	time and a half

What would be the weekly wage for this job?

4

[Turn over

DO NOT WRITE IN THIS MARGIN

Marks

13. PQRS is a rhombus.

The diagonals PR and QS are 15 centimetres and 8 centimetres long as shown.

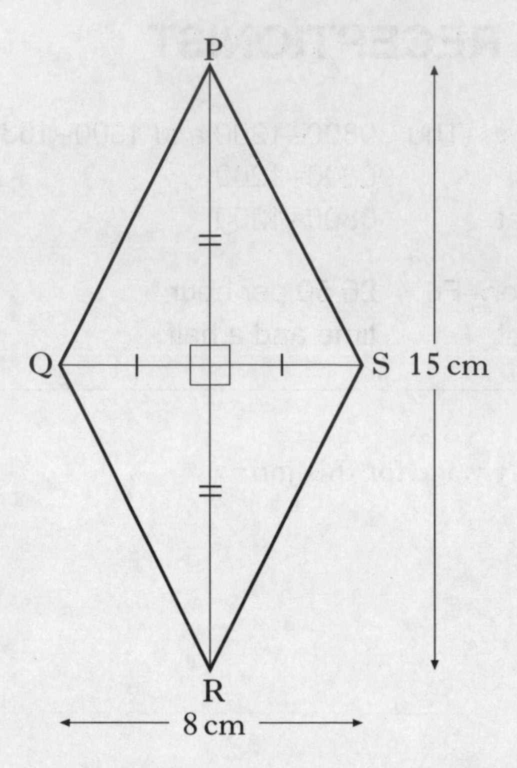

Calculate the length of side PQ.

Do not use a scale drawing.

3

14. Margaret is recovering from an operation.

She needs to take 4 tablets each day for a year.

The tablets are supplied in boxes of 200.

Each box costs £6·50.

How much does it cost for the year's supply?

3

DO NOT WRITE IN THIS MARGIN

Marks

15. The diagram below shows a plan of a patio.

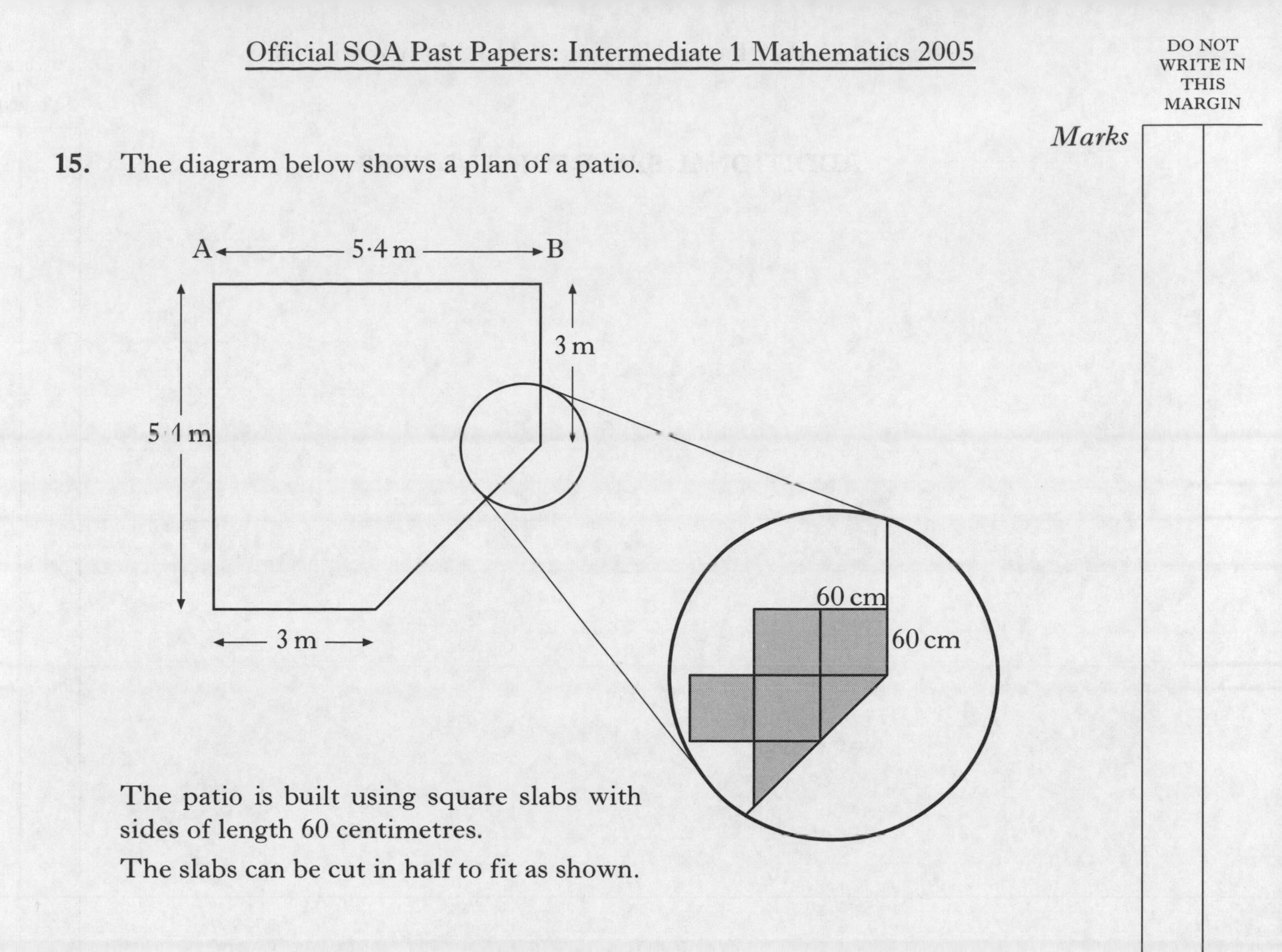

The patio is built using square slabs with sides of length 60 centimetres.
The slabs can be cut in half to fit as shown.

(*a*) How many slabs fit exactly along edge AB? **1**

(*b*) How many slabs are needed altogether to build the patio? **4**

[*END OF QUESTION PAPER*]

DO NOT WRITE IN THIS MARGIN

ADDITIONAL SPACE FOR ANSWERS

DO NOT WRITE IN THIS MARGIN

ADDITIONAL SPACE FOR ANSWERS

[BLANK PAGE]